Thomas Collins Simon

A Treatise on the Solar Illumination of the Solar System

Thomas Collins Simon

A Treatise on the Solar Illumination of the Solar System

ISBN/EAN: 9783744675383

Printed in Europe, USA, Canada, Australia, Japan

Cover: Foto ©berggeist007 / pixelio.de

More available books at **www.hansebooks.com**

THE SOLAR ILLUMINATION

OF THE

SOLAR SYSTEM.

What shall you think, my dear Kepler, of the leading men of science here, when I tell you that, like the deaf adder that stoppeth her ears, they will not look at the Moon and Planets through my telescope, although I have had it offered to them for the purpose a thousand times? They will not even look at the telescope itself. As the adder closes her ears so they close their eyes, to avoid the recognition of what is really going on.

GALILEO'S LETTER TO KEPLER, FROM PADUA,
August, 19, 1610.

[*** Quid dices [mi Keplere] de primariis hnjus gymnasii philosophis qui, aspidis pertinaciâ repleti, nunquam, licet me ultro, deditâ operâ, millies offerente, nec Planetas, nec Lunam, nec perspicillum videre voluerunt?—Verum ut ille aures, sic isti oculos contra veritatis lucem obturârunt.]

A TREATISE,

IN POPULAR LANGUAGE, ON

THE SOLAR ILLUMINATION

OF THE

SOLAR SYSTEM,

OR,

THE LAW AND THEORY OF THE INVERSE SQUARES;

BEING

AN ANALYSIS OF THE TWO RECEIVED LAWS

RELATING TO

THE DIMINUTION OF LIGHT BY DISTANCE,

WHEREIN IT IS SHOWN THAT, ACCORDING TO UNDISPUTED FACTS OF
NATURE AND OF SCIENCE, THE SOLAR ILLUMINATION IS
EQUAL THROUGHOUT THE WHOLE SYSTEM, AND
THE LAW OF THE INVERSE SQUARES FOR
LIGHT, PHYSICALLY IMPOSSIBLE.

TO WHICH IS ADDED

THE PROSPECTUS FOR A PRIZE OF FIFTY GUINEAS OFFERED FOR DISPROOF
OF THE SCIENTIFIC FACTS HERE FOR THE FIRST TIME INDICATED.

ALSO

AN APPENDIX OF EXTRACTS FROM THE WRITINGS OF PROFESSIONAL MEN.

By COLLYNS SIMON, Hon. LL.D., Edin.,

AUTHOR OF "SCIENTIFIC CERTAINTIES IN PLANETARY LIFE, OR NEPTUNE'S LIGHT AS
GREAT AS OURS" (1855.) "THE NATURE AND ELEMENTS OF THE EXTERNAL
WORLD," 'THE LOCALITIES CONNECTED WITH THE MISSION
AND MARTYRDOM OF ST. PETER," &c., &c.

WILLIAMS AND NORGATE,

14, HENRIETTA STREET, COVENT GARDEN, LONDON;
AND 20, SOUTH FREDERICK STREET, EDINBURGH.

1879.

TABLE OF CONTENTS.

THE FOUR PARTS.

PREFACE.

HAVING endeavoured, in short statements, without success, for nearly five-and-twenty years, to induce professional men to study the subject in Physical Optics here mainly treated of, viz., the Theory and Law of the Inverse Squares, and the Illumination of the Solar System by the Sun, I now propose to lay the facts more in detail before a more extended class of readers,—before the scientific who, not being Professors, may therefore have more time to attend to what is new, and before the more thoughtful of the community, even though these may hitherto have made no special scientific studies. In the interests of this latter class of readers, every statement here made will be found entirely divested of Technicalities,—those Technicalities which, however useful as stenography for Professors among themselves, I cannot but regard as the *closed doors* of Science, by which public discussion is consciously and carefully excluded, and by which, no doubt most unconsciously, scientific progress is impeded, as all progress is, and must be, in the dark corners of Monopoly and Protection.

It was in 1855, in reply to the Essay on the "Plurality of Worlds," by a writer of great distinction, that I published a small volume on some facts of Science connected with Life upon the Planets, for the express purpose of drawing attention to this equal distribution of the solar Light, as well as to some few other points of planetary science which also seemed to me to have been hitherto unaccountably neglected. That little work was examined, with their usual liberality, by our two highest authorities in the science of Optics,—Sir David Brewster and Sir John Herschel. They had the kindness to point out in the work an error respect ing the exhausted receiver, but neither of them found any error to point out in what was there stated respecting the distribution of the Sun's Light throughout the system, both merely saying, in kind and candid letters, that my general conclusion was so extensive, and involved so many considerations, that they could not, with their large amount of daily work, examine it sufficiently, but that they saw no reason to doubt that the ordinary opinion was correct. As, however, my conclusion does not in itself involve this necessity for any extensive studies, I supposed their meaning to be that it would require a great deal of investigation to discover whether it clashed, and how far, with any popular modern hypothesis, or how, at least, it was to be reconciled with such; and this, of course, I could not deny, might be the case; although, as far as I could judge, I was not able to see that any study whatever would be required for that purpose.

Many other professional men then, and afterwards, received copies of the work, and letters respecting it, but as is so usual in the case of men occupied in business, they were unable to take notice of either.

In the "Corrispondenza Scientifica" (which has been published during the last thirty years, on the Campidoglio in Rome), I wrote in 1875, at the desire of the Editor, an article on one portion of the same subject, viz., on the logical defect in the Law of the Inverse Squares; which article was subsequently translated into French, and much circulated among men of science in European countries. It produced, however, no discussion that I have heard of. I afterwards, in the hope of drawing attention to the subject, offered further articles to various English Journals suggested to me by the Professors whom these Journals rely upon in such matters; but in no case were the articles published, although offered gratuitously, nor the least notice taken of the question. In 1876, I presented a "Memoire" on that subject to the French Institute, at the suggestion of an Astronomer of European fame, as well as subsequently to the Lincei in Rome, at the suggestion of another Astronomer equally distinguished; also to the Istituto Lombardo at Milan, through one of its most celebrated members; but in all three cases without obtaining any investigation of the scientific facts I indicated. It is now by the kind advice of a foreign Physicist who deservedly takes rank among the most eminent in all that relates to Light and Optics, that I have written, with so much detail, the present pages, as, in his opinion, the most likely means of obtaining the discussion I so much desire. This effort, however, to drag into view the extraordinary errors so long and so dogmatically taught under a covering of technical language, I have not entered upon until I had found it impossible to induce anyone else, Professor or not, to undertake the task of doing so, whose name alone, already in favour with the public, might have rendered it so much more easy a task for him than it is likely to prove for me.

I proceed now to another department of what here requires to be explained. I have in these pages occasionally adverted in terms of discontent (or, as some will express it, of reprehension and remonstrance) to the utter disregard, manifested by the advocates of this Theory, for anything like scientific certainty respecting it, and to the discouragement of its discussion, on the part of those who could have best promoted such discussion, as well as to this unaccountable surreptitious recognition of a mere "opinion" as a scientific fact. But here it is my first duty to state that I have found nothing of all this in the case of unprofessional scientific men. I have found nothing whatever to complain of with regard to the attention given willingly to the subject by such men, nor to the fearless judgment which they all uniformly express respecting it; and even with regard to many Professors,—especially with regard to the more refined and distinguished of these, I have nothing to utter but gratitude for such little attention as their enormous professional work allowed them to afford me. Notwithstanding this, however, or rather on account of this, the distinction between scientific men who are professional and those who are not, is, upon this question, an important one. The latter have leisure for new subjects and for subjects which are unattended with financial advantage. The former have not. Those who live by the Profession and their fame have already so many subjects, lectures, articles, books, consultations, &c., on hand, that they have no time or strength for more. They are as it were in chains; and chained moreover to what is already before them. Besides which, there is an Orthodoxy in the profession of a Science as well as in that of a Religious form; from the bondage of which Orthodoxy the priests of Science are no more allowed, than other priests, to consider themselves exempt. For this reason the free scientific

man is more open to listen to new discoveries, and to investigate them, and to fling off mere Orthodoxy. Thus it
is that during that quarter of a century to which I advert
as having been occupied with this subject, I have rarely
found a scientific man exempt from the "business" of
Science,—in other words, I have rarely found an unprofessional scientific man,—who was not disposed to examine
these strangely neglected questions, which constitute,
really, the whole of one of the three great departments of
Physical Optics (viz., Photometry), and even, after the little
attention requisite, disposed to regard my facts and reasonings as being conclusive, *i.e.*, as being entirely exempt from
mere opinion; whereas in that time,—a long time now-a-
days in scientific progress,—I have found no Professor
who, to say nothing of any other hindrance, had even the
little *leisure* necessary to study these Propositions, and
comparatively but few who even had patience with the
unprofessional man who considered such questions his own
as well as theirs.

Such discontent then as, in these pages, I may be found
to express, must not be misinterpreted. It is not ever
against men of Science generally that it is expressed, but
only against Professors; nor against the profession generally, but only (1) against some of its members whose
idiosyncrasies I have had to encounter in my efforts to
bring these facts in the most effective way before the
public; and (2) against that lamentable incubus which
weighs upon the whole Profession of Physics, to whatever
source it is to be attributed, and for which its members do
not seem to be responsible, but which so restricts the action
of the more competent Professors that they have not the

least time free from teaching,—not even what would suffice
to investigate the truth of what they teach. It must not
therefore be supposed that I here blame professional men as a
body; I do not; however much I could animadvert upon the
conduct of individuals. The only two points which I bring,
in any sense, against professional men generally are two
which they will all, themselves, at once admit the truth of,—
even those who are the most opposed to my exertions,—
and which, at least on this question, they will now, I doubt
not, themselves speedily rectify; while, in contributing to
this rectification as far as lies in the power of one who is
not in the Profession, I cannot but consider that I am co-
operating with them,—certainly with the more enlightened
of them, instead of thwarting them. One of the two points
to which I allude is, that they all teach the doctrines shown
in these pages to be false, and which, if they could even
be only suspected of being false, no Lecturer would care to
teach. The second is, that they have invariably declined
to examine into the truth of these doctrines, or of the facts
which contradict them, whenever applied to for that purpose. The Profession, generally, will do me the justice to
recognize that, in what I have here written, these are the
only two statements I make respecting them, and that these
two statements are true. However curious such circumstances may be, and however much to be regretted, there
is not much to blame individuals for in either of them;—
not in the first, because we are all liable to make mistakes,
especially when we indulge in traditional science, taking
things on trust from our ancestors; and not much to blame
in the second of these circumstances, except only where
investigation is possible yet neglected, or where it is refused

either with childish fretfulness or under some mistaken notion of professional infallibility.

One of the two statements here in question, is that all the Professors teach, for instance, the five following Principles :—

First, that Light *spreads*, thus thinning out and becoming weaker ;

Secondly, that we have 900 times more Light from the Sun than Neptune has ;

Thirdly, that at any distance whatever from a luminary, the Light is four times greater than at double that distance ; —in other words, if we take any length in the line of distance from the source, the Light at the beginning of that length is always four times greater than that at the other end of it ; and this, whether the length taken be an inch or a mile ;

Fourthly, that the light of 500 candles would not require more medium to absorb it, than the light of one candle would require ;

Fifthly, that no amount of Medium, however extensive or however dense, can ever wholly absorb even the light of a single candle.

Here then is one of the only two points on which I can be said to remonstrate with the Profession, if indeed my words even amount to a remonstrance. Why have these doctrines, I seem to ask, been so long taught in the Royal Society and at the Royal Institution? Ought we not to be a little ashamed of ourselves that such an extraordinary incident should have occurred among us—such strange credulity have grown up in our honoured halls of science?

and does it not justly excite the fear that other scientific errors of equal obviousness and equal magnitude may remain behind? But surely with regard to such Propositions as those just cited, and to this general remonstrance, if such it is, respecting them, I cannot well be found fault with until the ordinary teaching is shown to be correct, and my facts and reasonings shown to be erroneous.

The other point which I bring against the Profession as a body, and which they will recognize not only as being the only other point which in these pages I do bring, but also as being something fairly and justly attributed to them, is that whenever the investigation or discussion of these five principles has been either publicly or privately proposed to them, they have not only declined it, but discouraged it. This they will all admit. It is, moreover, clearly proved by the fact that although, in one form or another, the question involving these principles has been now before them for nearly five-and-twenty years, and urgently—perhaps it will be said importunately—pressed upon their attention, not one of them has yet undertaken its discussion.

Now, as to the truth of these two statements,—call them allegations if you will,—there can be no dissension between me and the Professors,—the five principles in question being universally held and taught by all of them, and all, as a body, resolutely refusing to examine them;—the more enlightened being able to combine with the strangeness of this refusal nothing but what good nature might dictate to them, and the less enlightened not, of course, even that;— the more cultivated and enlightened refusing with kindness and consideration, declaring their inability through want of leisure to study the question, or their unwillingness, through

conviction, to encourage its discussion; while the less cultivated refuse, in the usual coarse way, either by ignoring
all existence of scientific knowledge outside the Profession,
or by merely taking no notice whatever when addressed
upon the subject, or by manifesting a curious kind of irritation,—almost resentment,—when it is suggested, as gently
as possible, that perhaps the existing opinion on this subject
might be found erroneous;—acting thus, as if conscious of
some intellectual incapacity to discuss such questions, or
even to utter a word respecting them, and as if silence,
irritation, or self-complacency were the only means whereby
they could hope to conceal this incapacity from the stranger.
In only two or three trifling instances did this evasion take
the form of impertinence, and of that "insolent admonition"
which, under similar circumstances, Dr. Young had also to
complain of from the ignorant. Of the very young and
inexperienced, a few seemed to think that they sufficiently
"protected themselves" by recommending me to procure
and study some book they had lately written, or to study the
Polarization of Light, or the Conservation of Energy, &c.,
as all offering "views" adverse to *my* "views;" while others
of them at once retreated, each into his own little im-
promptu forest of algebraic symbols and geometrical diagrams, out of which they did not afterwards venture to
emerge. So senseless, however, and unintelligible has all
this singular conduct frequently appeared, as to give an impression that the applicant was suspected of endeavouring
to discover some *professional incompetence*, and of seeking to
supplant the occupants *in their chairs;* to some alarm of
which kind upon their part, I have been compelled to attribute not only much of the unscientific proceeding I advert

to, but also the remarkable fact that I have *generally* found
Astronomers to be more ready to speak, and to listen, on
the neglected questions of Photometry here under consideration, than I have found the mere Physicist to be; as
I have also found *always* the unprofessional man of science
more disposed to do so than the mere Professor.

But while thus endeavouring to specify, with as much
precision as possible, the experience, which during all these
years, I have had of the professional world upon this neglected branch of Optics, and to explain the grounds of my
dissatisfaction with a considerable number of the Professors,
I have great pleasure in stating that, although I have
received from none what could well be called encouragement, I have without exception received from all the more
eminent in the professions of Physics and Astronomy
that kind and often generous attention which belongs as
much to the love of Science as to the courtesies of life, and
which enabled these distinguished men at least to see and
to say that the proposed investigation would involve more
study than they had time for at their disposal; or, as some
few of them in this country preferred to express it, that,
although what they taught was mere matter of opinion,—
what therefore they could not establish by experiment or
by proof, yet they were so "completely convinced" of its
truth as to regard any investigation of it but as time mis-
spent and withdrawn from more profitable occupations.
And among those enlightened men of whom I speak, I may
be permitted to mention the world-famous names of Respighi,
Schiaparelli, Tacchini, and the late Father Secchi, in Italy;
Fizeau and Janssen in France; Helmholtz and Lommel in Ger-
many; with Brewster and Herschel among ourselves; as well

as Professor Stokes, Professor Tyndall, Mr. Proctor, and Sir William Thomson. I repeat, however, and wish it to be distinctly understood that, from none of these distinguished men have I received the slightest encouragement in the task I had set myself. Nor is that all. Although none of them have pointed out a single fact or reasoning of mine which they could speak of as inaccurate, only two or three of the gentlemen I have named, and those foreigners, could recognize even the *possibility* of my conclusions being correct; some of the others, however, not unnaturally perhaps, misinterpreting these conclusions, and finding fault, not with what I said, but with what they, from the cursory glance which alone they could afford, supposed that I probably intended to say. To some of the most eminent of our professional men, in consequence of their advanced age, and from unwillingness on that account to give them trouble, I have not applied for their co-operation in this matter, well assured however, that if I had done so, I should have received from them the same courtesy and kindness as from so many others, although probably the same discouragement as from all.

. It will be seen, then, from the foregoing circumstances, why I now seek a more extended auditory, and that this is done without the slightest sense of opposition to the wishes of the great leading lecturers of the day, but rather with the purpose of co-operating with them, in my humble way, by promoting that discussion and investigation about the success of which they experience in themselves so much discouragement or which they are precluded from promoting, in consequence of the large amount of professional work which they already have on hand. In fact it is, as 'I have already mentioned, only after this step of a detailed exposition upon my part was good-naturedly suggested to me by one of their most eminent colleagues, that I have taken it, and have thus embarked in this attempt to make intelligible to all readers the extensive principles of Physical Optics here for the first time treated of, and as I am told for the first time discovered. I have, however, taken this step not merely to arrest the attention of scientific men generally, and even of the unscientific reader, but especially in the hope of bringing all who teach in Astronomy and in Physics, —both the more enlightened and the less enlightened,— round at last to give a little more attention to the Theory of the Inverse Squares in their Lectures and their Articles; and, when they see the need, to correct with all the energy necessary, the mistaken notions resulting from this theory, and now so current respecting (among other matters) this unequal illumination of the solar system by the Sun. It is also with this same object, and in the hope of enlisting the sympathy of professional men everywhere in some discussion of the subject, that the PRIZE OF 50 GUINEAS is here offered for disproof of the natural facts here pointed out and acknowledged by all to have been hitherto completely unknown,—even unthought of,—so completely unknown and unthought of that, in the eyes of professional men, these facts have now all the appearance of being impossibilities; this disproof to have, as may be seen in the annexed Prospectus, exclusive reference to the judgment of those who are in the capacity of public Professors either in Astronomy, in Optics, or in general Physics; and either in Europe or in the United States of America as well as in any of our Colonies.

It will not be out of place to advert here to the fact that the present is by no means an exceptional or isolated instance of the reluctance with which each individual Professor of Physics gives his attention to innovations or discoveries, when they do not proceed either from himself, which of course cannot often happen, or from some member of greater authority in the Profession than himself, (for authority with the profession of Physics has as much weight as in the Church and in the Law), or when a practical application or public excitement, called " pressure from behind," does not invest the new idea with some pecuniary interest ;—a reluctance this,—whether merit or defect, which is as I say, considerably increased in its operation, when the discovery originates with some one either wholly or mainly disconnected with the Profession. It is not reluctance only that is then experienced. It is with distinct repugnance that each member of the Profession turns away from the studies of those who " work as original investigators,"—the teachers of the teacher, and who are not paid for the instruction which they furnish, who occupy themselves with Science *con amore*,—from love of Science only, not for gain, and who are therefore called, by the Professors, " Amateurs," a term now of ridicule if not reproach, in spite of all that, for another purpose, has been *said* by some Professors to encourage scientific studies outside the Profession, (*i.e.*, studies exempt from mercenary motive) in those who have the disposition and the leisure for them ; or, as it has been expressed, " to develop and deepen sympathy between Science and the world outside of Science, deeming it good for neither world to be isolated from the other, or unsympathetic towards the other."

The present is by no means a solitary case of this repugnance. The feeling in question would be found in such cases, on the contrary, to be the ordinary state of things. To mention but one of the many instances upon record, and to omit all disputes and discouragements of mere Professors among themselves, I may remind the reader that, in the beginning of this century, Dr. Thomas Young, the distinguished physician alluded to in a former page, although his innovations were afterwards most enthusiastically adopted by all the Professors of Physics, was for twenty years unable to prevail upon Sir Humphry Davy and the other leaders of Physical Science at that time in England, to attend to the very small amount of reasoning by which he was enabled perfectly to establish his statements. They not only refused to accept his conclusions, but even to attend to his reasons for them. They all represented themselves as either " quite convinced" or " sufficiently convinced" that the ordinary " view" of matters,—the ordinary " opinion,"—was the correct one;—and this on no other intelligible grounds than because he did not belong to the Profession, and because the subject which he urged upon them, and which they had overlooked, was one of mere reasoning instead of mere experiment. For it cannot be supposed that they were unable to understand him until later, nor can it be alleged, in palliation of such conduct, that there was, at the time, any *primâ facie* improbability of his having fairly arrived at the conclusion which he undertook to explain. On the contrary, he was known to be eminent in his own most difficult Profession (for which perhaps he was in the habit of expressing too strong a preference to please them), and is admitted by all to have been endowed with the energy and the intelligence which are, when combined, so easily mistaken for that special nature of mind which we call " genius,"—a term now often applied to him.

During all that time, then, this remarkable man was sedulously " pushed from the public mind," as it has been well expressed,—" quenched for twenty years,—hidden from the appreciative intellect of his countrymen," and treated as " a dreamer," the papers he presented, from time to time, to the Royal Society, being wholly disregarded, and all by the very men whose duty it was to have helped forward the discussion which he asked for, *whether it seemed of use or not to do so*, and even if his reasonings, instead of being short and simple, had been extensive and complicated. It has been justly said that " there is no calculating the amount of damage these twenty years of neglect may have done to Young's productiveness as an investigator ;" and again : " If you starve or otherwise kill the scientific discoverer, (in those studies whose importance you cannot understand,)—nay, if you fail to secure for him free scope and encouragement,—you not only lose the motive power of intellectual progress, but infallibly sever yourselves from the springs of industrial life." The following passage also, from the same pen, has a strict application to such a case as that of Dr. Young and the Professors of Physics in his day :—" Let me remind you that the work of the Lecturer is not the highest work ;—that, in Science, the Lecturer is usually only the distributor of intellectual wealth amassed by better men. . . . You have scientific genius amongst you,—scattered here and there. Take all unnecessary impediments out of its way. Keep your sympathetic eye upon the originator of knowledge. Give him the freedom necessary for his researches, not *demanding from him practical results,*—above all things, avoiding that question which ignorance so often addresses to genius, ' *What is the use* of your work? What *could* be the use of your discussing it ?' Let him make truth his sole object, however unpractical, for the time, and unprofitable the investigation may appear to you. Never mind the *use* of it. Give it what encouragement you can ; and if you have this confidence,—if you give the student courage and give him sympathy, although you understand him not,—if you thus, in the Eastern phrase, cast your bread upon the waters,— be assured that you will find it again, though it may be after many days, in substantial and practical profit, both to your industries and to your science." [*The latter clause of this eloquent address has been here slightly paraphrased from the original, in order to prevent it from being misinterpreted as an appeal in aid of funds for the Profession, which, of course, was not intended where this passage occurs.*]

A few writers of eminence,—themselves Professors of Physics,—seem under the strange impression that it was not the members of the Profession, but unprofessional people,—nay, one unprofessional and very young man,— who refused the discussion of Dr. Young's propositions, and even declined to read his papers, thus prejudicing the professional class against him, and securing his condemnation by the only tribunal before which he could have been tried ! But this is an obvious misapprehension or forgetfulness of what occurred,—a misapprehension or forgetfulness which it is extremely difficult to account for, unless we suppose it produced, to some extent, by the sense of shame which the odious fact in question could not fail, in these days, to excite in the mind of any of our more enlightened Professors. Unprofessional people neither prevented the professional Public from examining and discussing Dr. Young's arguments, nor would it have been possible for a whole nation of them to have done so. We all know very well, and experimentally, that no unprofessional man could prevent a proposition from being discussed or held by our leading scientific teachers if these chose to discuss it or to hold it, however preposterous such a proposition might be, nor even prevent it, if approved of and encouraged by them, from being fully recognized, or at least discussed, by the unprofessional. We know indeed that one Professor of Physics will often succeed in preventing the conclusions of another Professor from becoming popular even in the Profession, or the conclusions of an unprofessional man of science from being attended to by other Professors; but how can any one suppose that the paid teachers of Physical science, at the head of their Profession in the beginning of the present century, would have been deterred from discussing Dr. Young's arguments merely because some *one* unprofessional writer, — anonymous moreover, — said *he* thought them ridiculous, and what the Royal Society ought not to sanction. This would be giving but a poor account of the Royal Society, and a most miserable estimate of our professional community, to say nothing of the circumstance that what really occurred was entirely different, and is now well known to have been so. But, independently of our biographical records, we have abundant evidence of the apathy, if not antipathy, with which the Professors received Dr. Young's innovations, even if we go no farther than the three facts admitted by all writers on this subject; viz. (1) that there is no trace of any answer, by a Professor, to the youthful satirist in the *Edinburgh Review*, except that written by Dr. Young himself; (2) that, of Dr. Young's answer not more than *one* copy was sold in England; and (3) that there was no serious examination or discussion of his propositions by English Professors until they were already fully recognized in France as true. Surely there can be no doubt in any one's mind of the neglect and discouragement here in question. Dr. Young was "too strong (it has been truly said) to be tied down by professional regulations," and "he worked as an original discoverer;" which bold proceeding the Professors of Physics only tolerate, in their colleagues, when even in them, discouraging it strongly in Physicians and all others, notwithstanding the equivocal overtures of a contrary character, already alluded to as occasionally made to "the Public" by one or two of our leading Lecturers.

For twenty years, then, he effected nothing, "overborne by the authority of his antagonists." Although "an investigator of Truth for Truth's own sake," in the truest sense of these words,—seeking no pecuniary advantage from his work,—he found neither "promoters nor protectors" among those who, better than any one else, could have "supplied his discovery with the freedom, light, and warmth so necessary for its development;" and we read in the sad annals of his story, that this "Amateur" of the Professors never received from any of them,—even from those among them who were his personal friends,—the slightest co-operation nor encouragement, until a generous foreigner, a young Engineer Officer of spirit and ability, devoted to optical research, and whose first *Memoire* on Light had just been crowned by the French Institute, gave these personal friends and the rest to understand, even in that *Memoire* itself, and before the world, that there was something for scientific men to be ashamed of in their conduct to Dr. Young;—that what this distinguished man had tried to teach them, was what everybody could easily understand, even if everybody could not have thought it out; that he had once supposed himself the first discoverer of it, but that he now fully recognized Dr. Young's prior claim to that distinction; and that this discovery was what thenceforward every intelligent Lecturer would have to hold and to teach. Hereupon the professional members of the Royal Society began, for the first time, to study the statement of the English Physician as printed in their " Transactions," and, " startled at last into a consciousness of their injustice to him," vied now with each other in throwing themselves, midst regrets and compliments, at the feet of him whom they had wronged. They made him their Foreign Secretary by acclamation, and were, many of them, intent upon making him their President.

Thus after twenty years of this scientific exile,—twenty long years passed in " the hampering toils" of this obstruction and this discouragement, without being able to obtain the least attention to his discovery, " having great names arrayed against him,"—(that professional obstacle to progress,) the obstructive " authorities" were at last, in their turn, overborne by that " pressure from without," which never fails. They could not afford to be behind the Professors of the Continent; and the reasonableness of what Dr. Young insisted upon was fully recognized, first indeed, as I have said, and as generally happens, by foreigners, but afterwards recognized, to such an extent, by English Professors themselves, that almost all of them have since rushed into the opposite extreme of the wildest admiration for everything respecting him. One great admirer evinces this enthusiasm as follows :—" To give you a notion of the magnitude of this Man, let Newton stand erect in his age, and Young in his. Draw a straight line from Newton to Young, tangent to the heads of both. This line would slope downwards from Newton to Young, because Newton was certainly the taller man of the two. But the slope would not be steep, for the difference of stature was not excessive. The line would form what Engineers call a gentle gradient from Newton to Young. Place underneath this line the biggest man born in the interval between both. It may be doubted whether he would reach the line; for if he did, he would be taller intellectually than Young, and there was probably none taller." The reader will not be surprised to hear that this was written by an eminent member of the Profession which had been so unjust to Science as well as to Dr. Young.

All this discouragement, obstruction, and neglect to which I have adverted, on the part of the Royal Society, and of the Professors to whom Dr. Young's innovation had so unnecessarily appeared unintelligible (if that is the excuse which we are to offer for their treatment of him), as well as the full appreciation of his merits, in the subsequent generation both at home and abroad, are in the following passage, concisely borne witness to by the amiable and accomplished Helmholtz :—

" Dr. Young's was one of the most profound minds that

the world has ever known; but he had the misfortune to be too much in advance of his age. He excited the wonder of his contemporaries, who, however, were unable to follow him to the heights at which his daring intellect was accustomed to soar. His most important ideas lay, therefore, buried and forgotten in the folios of the Royal Society, until a new generation gradually and painfully made the same discoveries, and proved the exactness of his assertions and the truth of his demonstrations."

It cannot be denied that there is something no less appropriate, as instruction, than as illustration, in the citation of these facts.

No doubt there will be the necessary critic to say that I am seeking to compare myself to Dr. Young; but I am not. I am comparing my case to his, not myself to him; and the cases resemble each other in many remarkable particulars, four of which I may here mention:

1. What has been, in either case, contended for, is neither an hypothesis nor a theory, as some have imagined it to be, but a natural fact which seems in neither case to have been discovered before;—in his case, the interference of water-waves; in mine, the non-deviation of the ray of Light from its straight line,—the fact that it does not spread, thus thinning out and becoming weaker, upon one size of surface more than upon another.

2. These two natural facts have been, in their respective cases, wholly discredited by *all* the English Professors connected with the subjects, who have not only refused to discuss the facts, but moreover even to attend to the evidence of them, on the ground of its being "of no use to do so," as the point at issue was, in both instances, a matter of

mere " opinion," and one on which these Professors had
" already made up their minds."

3. Dr. Young had two aims,—two propositions,—both
controverted; I also have two, and both controverted.
He was defending an optical hypothesis, the wave-theory,
denied by most people in his day, and, in support of it,
defending also, at the same time, this obscure but curious
natural fact, the interference of waves in water, which was
not only denied, or at least neglected, by most people, but
by everybody. I am pointing out that an Optical theory,
asserted by all,—the theory of the Inverse Squares, is
utterly and manifestly false; and, in order to make this
clear, I seek to explain that Light from its nature does not
and cannot deviate from the straight line as that theory
supposes that it can, a natural fact this also as stoutly
denied or neglected by all, as the interference of waves can
well be said to have been.

4. The refusal of the learned to study the fact extended
obstinately, in Dr. Young's case, over twenty years, when
at length the truth of his discovery was recognized abroad;
and, in the present case, already over nearly five-and-twenty
years, with, as yet, but a very little glimmering of this
recognition; and that little, as usual, not in England.

In two other particulars our cases differ;—these points
of difference, however, so neutralize each other, that the
obstacles to the triumph of truth have been, in both cases,
pretty much the same. The success of the first case, therefore, seems to augur favourably for the second:—(1.) Dr.
Young's fact (the principle of wave-interference) was more
obscure and more difficult to be explained than mine, and,
perhaps, it will be admitted, less urgent when he sought to

draw attention to it, as well as less important; for it was chiefly of use as being supposed to favour the undulation-hypothesis, then in little repute; all which must have constituted an obstacle for him, that does not exist for me, inasmuch as, in my case, it is only for the wholly unscientific reader, if for any one, that the natural fact can be attended with any difficulty; and only by him, if by any one, that it can be regarded as, in any sense, uninteresting and unimportant; since upon it there depends the truth or falsehood of the Inverse Squares in Optics, and the equality or inequality of the Solar Illumination throughout the system. (2.) The second of these two points of difference is this: There were a few foreign Professors, advocating, on the same occasion, and with considerable energy, the same hypothesis as that in support of which Dr. Young put forward the curious fact of nature to which, as well as to the hypothesis itself, all in England refused the least attention. These few foreigners had therefore some little interest in this alleged fact of nature, and in its being successfully pointed out; whereas this advantage I have not had at all. There is no question pending in the professional world that could give the great fact of nature which I point out, any special party interest there (the interest of all being, on the contrary, "to push it aside" and upon that point to have no discussion)—nothing, therefore, offering hope of any influence except mere brute "pressure from without," "to break through the circle here drawn, for so many generations, around the operations of the human intellect."

This is a considerable and rather discouraging point of difference between Dr. Young's case and mine. The dis-

tinguished Physician was, it is true, engaged in defending
a proposition in which no one, in England, then took any
interest (viz., the wave-hypothesis for Light), and was doing
this through a physical fact, denied then, or rather un-
considered, by all except himself (viz., the interference
or combination of waves in water); but while he was, in
England, in this hopeless condition of one crying in the
wilderness, there were already in other countries some
influential Professors zealously engaged, they also, in
search of every indication that could be discovered in
favour of the same undulation-hypothesis, although entirely strangers to the natural fact proposed by an English
Physician in support of it; — persons, therefore, who
were not likely to leave unturned the stone under which
this Physician told them they would find all they wanted.
This co-operation, such as it was, I have not had. I seek
to make known, to a singularly reluctant auditory, as far
as the professional world is concerned, my great fact of
nature (viz., that Light does not spread from its straight
line, and thus diminish in intensity,—that the theory of the
Inverse Squares is therefore false, and that the solar system
is all equally illuminated by the Sun), and this I do without any such aid as Dr. Young's fact, derived from the
existing pursuit of an hypothesis ;—mine being, on the
contrary, a fact which not only furnishes no proof of anything at present in demand, but which can even be, as it
has been, overlooked by all, and trampled on by all, in their
headlong pursuit of far less important though more curious
discoveries. I cannot boast of one single teacher of Optics
or of Astronomy, even slightly interested in the truth of
my fact, either in England or out of it. The only Professor,

—but he, of universal renown,—who, in his conversations with me, had the courage to say,—and said it frequently,—" Perhaps you are right," is no longer living, and I much doubt whether he has left any record even of his imperfect appreciation. But this great man, as the word " perhaps " here implies, had not the leisure to examine the subject further, weighed down as he then was, both with years and work. I am thus at this hour, as for so many years, wholly without co-operation in these efforts to make known the obvious fact which I have explained in the present treatise ; and this utter isolation seems to counterbalance the advantage connected with the other point of difference between Dr. Young's case and mine, leaving the difficulty, as I have said, pretty equal for both of us. But this difficulty, History thus teaches us, is not insurmountable. The enlightened men who fill our chairs of Optics and Astronomy are not now more completely " convinced," that the Optical theory of the Inverse Squares is true, that Light can spread, and that vast portions of the solar system have billions of times as much Light as other vast portions of it, than in Dr. Young's time, they all were, that the undulation-hypothesis for Light was false, and that his " innovation," in support of it,—the interference of Water-waves,—was the " pitiable nonsense " it was described to be in the *Edinburgh Review.*

Although, therefore, my controversy has been always, as it still is, single-handed against the Profession, and has thus been, naturally enough perhaps, more prolonged than Dr. Young's, it can be hoped that it now approaches its termination, and that this irresistible " pressure from without," which we are taught to consider as, in such a case,

the only moving power in the professional regions, is at last
about to break through the circle here also drawn for so many
generations around the operations of the human intellect.
" No authority," it has been truly and nobly said, " however
high, can long maintain itself against the voice of Nature,
when she speaks through experiment as well as common
sense;" and we need not despair that this theory of the
Inverse Squares and of the ray that spreads, will, sooner
perhaps than some imagine, share the fate of that hypothesis so long opposed without success by Dr. Young,
and now, for a time at least, supplanted by its rival;
especially as, in the case before us, it is not proposed to
replace the theory of the Inverse Squares by any rival
theory, but only by the plain and simple facts of nature
which it contradicts.

As the Optical theory of the Inverse Squares is not only
wholly false but also very generally misunderstood, even
by those who write or lecture on the subject, there have
been here two sets of errors to be pointed out instead of
one set of these. It has been necessary to explain fully,
as is done in PART I., the law and theory themselves, and
the misconceptions current respecting them, before proceeding to point out, in PARTS II. and III., the utter impossibility of the law as well as of its theory. PART IV.
relates merely to the conditions under which the Solar
Light is diminished throughout the system, and under
which alone all Light is diminished, when diminished by its
distance from the Source.

PROSPECTUS OF THE CONDITIONS,

under which, for the purpose of promoting discussion, a PRIZE OF FIFTY GUINEAS is offered by the author, for the disproof of those facts and reasonings whereby in this Treatise it is shown that the whole theory of the Inverse Squares is an error in Optics, and that the Solar Illumination of our system is so equally distributed that we should be unable o discern any difference in this respect between the different portions of the system.

1. The Prize is to be open for one year from the date of this Prospectus.

2. It is open to ladies as well as to gentlemen; and to the professional as well as to the unprofessional.

3. It is open to Europe and the United States of America; also to Canada, India, and all the British Colonies.

4. Each Essay is to have in writing, and with their respective signatures, the testimony of three Professors, selected by the candidate himself, or herself, from the professions of Astronomy, Optics, or general Physics, indiscriminately, to the effect that the Essay is, in the opinion of these gentlemen, the valid disproof required, in which testimony the separate propositions controverted as well as the candidate's arguments adduced against each, shall be distinctly stated, and each Professor's distinct affirmation of their validity appended to each of the candidate's arguments.

5. The gentlemen selected by the candidates to declare

d

the scientific accuracy of their facts and arguments must be actually in possession of professorial chairs in Universities or National Colleges; yet if declared in writing by such Professors to be competent judges, the testimony can be accepted of gentlemen sufficiently known who *have* formerly filled such chairs, or who have even merely written books on Optical or Astronomical subjects.

6. Any of these gentlemen may be selected in any of the countries to which competition for this Prize is open; and there is no objection to remuneration being given, when it is necessary to do so, for the time and trouble of writing a testimonial.

7. No two Essays may bear the signature of any of the same Professors, or of others attesting the scientific accuracy of the disproof in question.

8. All Essays duly attested, as above indicated, to be sent in to the proposer of the Prize, through the publishers of this treatise, before the end of the year during which the Prize remains open.

9. The Essays will then be immediately submitted to some one gentleman of the highest scientific position that can be prevailed upon to arbitrate, selected by a majority of the candidates, and with whom it will rest to determine not only which Essay is the best as disproof of the propositions in question, but also whether any of the Essays constitute this disproof.

10. In case there is only one Essay sent in, or no majority of the candidates attainable, then there shall be three arbitrators chosen instead of one, to decide by their majority; and the choice of these will rest with the proposer of the Prize.

11. The proposer of the Prize and author of this treatise, hereby pledges himself to pay immediately the Prize of Fifty Guineas to the candidate whose disproof shall be declared the best.

12. To render the precise subject of the Prize as clear as possible, it may be useful to specify that the two main propositions for the disproof of which this Prize is offered, are (1) that the illumination of the Solar System by the Sun is so equal outside the atmospheres of the Planets, that no difference which exists could be discerned by the senses; and (2) that the law of the Inverse Squares in Optics is utterly false; without here raising any question as to this law applied to Gravitation. It is well, also, for the same reason, to specify that the subordinate propositions by which the truth of these two main ones is established, and the disproof of which, or their disconnection with the main ones, it is also, therefore, indispensable to make quite clear, are (1) that Light of its own nature, does not, and cannot deviate from the straight line; (2) that it cannot therefore diffuse itself or spread without a medium; (3) that it does not consist of spokes (of Light) with un-illuminated intervals between these spokes; (4) that it does not, therefore, become even relatively less in the larger regions of space; (5) that when Light proceeds from a central source, *i.e.*, from such a source that each point of the surface illuminated is equally exposed to the whole source, then the whole Light of the source exists upon each point of the surface; (6) that there is, therefore, the same degree or amount of Light upon each *part* of the surface and upon the *whole* of it, since there cannot be more than the whole Light upon the whole surface; (7) that Light therefore is not inversely, as the areas, upon each

equal portion of them, or, as it is technically called, upon each unity of surface; *i.e.*, Light does not decrease as the square of the distance increases; (8) that we, therefore, have no reason for supposing any other diminution of the Solar Light outside our atmospheres, except that resulting from the Ether or medium through which the Planets move; and (9) that even if any such other sort of diminution could be shown to exist, we have none of the data or knowledge necessary for framing the law of the Inverse Squares respecting it, either in the ordinary interpretation of this law, or in any other.

July 1, 1879.

ON

THE SOLAR ILLUMINATION OF THE SOLAR SYSTEM;

OR, THE

LAW AND THEORY OF THE INVERSE SQUARES.

In Four Parts.

THE SOLAR ILLUMINATION OF THE SOLAR SYSTEM;

OR, THE

LAW AND THEORY OF THE INVERSE SQUARES.

𝕴𝖓 𝕱𝖔𝖚𝖗 𝕻𝖆𝖗𝖙𝖘.

PART I.

PRELIMINARY INFORMATION ON THE QUESTION AND ON THE
TWO WAYS IN WHICH LIGHT IS SAID TO BE DIMINISHED
BY DISTANCE; WITH EXPLANATION OF THE TWO LAWS.

FIRST SECTION.

THE QUESTION STATED.

THE common proposition on the subject now before us is, that the solar system is very unequally illuminated by the sun's light,—that many parts of it are millions, nay, billions of times less lighted than other parts of it,—that the planet Neptune, only thirty times farther from the centre than ourselves, has a degree of the solar light which is a 900th part of the degree that we have; and that the same ratio of diminution is true of all Light independently of any further diminution resulting from the medium through which it passes.

The proposition to which I now invite attention, and which is easily explained even for unscientific readers, contradicts this utterly. It is that the whole solar system is equally illuminated,—that the extremely minute differences in the solar illumination between the inner and outer portions of the system would, if we could make the comparison, be found to be entirely imperceptible to the human eye, and that the outermost and innermost planets have the same degree or amount of this illumination.

In order to make my proposition quite clear, it will be sufficient that I should confine myself wholly or mainly to that Law for the Diminution of Light by Distance, which is known as the Law of the Inverse Squares, and to the supposed facts of Nature upon which that Law professes to have been founded; for, although there is another diminution of Light always going on, and another Law to be considered, yet it will be seen that it is this Law and Theory of the Inverse Squares which alone has led scientific men to suppose the solar system so unequally illuminated,—an alleged inequality of Light which, as will be fully explained further on, is very much greater than people commonly suppose,—so enormous, in fact, that very few except the initiated can have any notion of what it amounts to, and but few, it would appear, even of these. I undertake in a few untechnical statements, to make it quite clear, for the understanding of all educated people, that there is nothing of this kind in nature; that, even according to the Theorists themselves who teach and defend this Law and Theory, there is no diminution at all possible in the case they speak of; nor (what is still more curious) the slightest pretext for supposing that there is any; and that even if there had been any of the diminution which their theory supposes, we could have had no knowledge of it, nor any data, nor means of arriving at any *data*, for ascertaining the Law respecting it which they thus promulgate.

The two points therefore to which I shall here mainly require to draw attention are—(1) The Physical Impossibility of the Theory itself, *i.e.*, of any diminution whatever in the alleged case, even upon the showing of the theorists themselves, and (2) The Impossibility in point of Logic, which exists in the Law that has been assigned to this diminution of Light, even if any such diminution had existed as that which they here assume. The explanation of these two points under their separate heads, and in the simplest language, with special reference to the Illumination of the system, will set the whole question in the clearest form before the reader; and, I need hardly add, that upon this point, what is true of the solar light is true of all Light.

SECOND SECTION.

TWO MODES OF DIMINUTION IN LIGHT BY DISTANCE SUPPOSED TO BE ALWAYS SIMULTANEOUSLY GOING ON IN NATURE.

I HERE write, mainly,—I may say exclusively,—for the general reader. It is therefore proper to inform him that there are two ways, in both of which, at the same moment, Light is supposed by Professors of Physics to be diminished in consequence of its distance from its source,—two totally distinct and different principles upon which all scientific men, or almost all,—certainly all writers and lecturers,— consider that the whole degree of Light, proceeding from any central source, undergoes diminution simultaneously, in its passage, according to its distance from that source.

One of these two ways is the well-known one of Absorption by the Medium. The other diminution, known only to scientific men, is entirely independent of all Medium,— supposed to take place merely in consequence of the greater area over which the same degree or amount of Light is spread, or expands, as it is believed, and thereby becomes diluted or weakened, without any reference whatever to a medium; the greater area of the object being vaguely supposed to be occasioned by a greater distance from the source of Light, whereas it will be seen that it has this size in consequence of its *geometrical* distance, *i.e.*, of its greater distance, not from the source of Light but from the apex of an hypothetical cone or pyramid.

All Light passes through some Medium, whether fog or

the interplanetary ether, or glass, or water, or mere air, or other transparent substance, (we never know it except after passing through some such medium,) and is thereby more or less diminished according to the amount of Medium it has traversed, and more or less rapidly, according to the greater or less density of the Medium. This is the diminution of Light with which we are all of us familiar; which is therefore no hypothesis and presents no difficulty.

The other diminution of Light,—the hypothetical one,—the one known only to the scientific, and which is believed by them to take place quite independently of all medium and of all absorption, is supposed to be occasioned solely by the enlargement of the area or space illuminated,—nay, to be exactly in proportion to this enlargement;—*this is the main point to be here attended to;* so that, *cæteris paribus*, the larger the area is, the less it is illuminated; *i.e. either* the Diminution then becomes greater, *or* the Light itself becomes less; two very different statements it is true, but the difference is here unimportant, as it is this latter interpretation which is always here assumed by the scientific as the correct one, and which therefore we shall adopt;—*the Light itself becomes less.* An area which is *twice* the size of another receives, it is thought, upon each spot, from the same source, only *half* the illumination which the other does, and an area three times the size of another receives thus only one-third of the illumination which that other receives; or, to say the same thing in other words, an area or surface half the size of another area or surface, receives twice as much light, on each square inch, as the other surface does; a surface three times smaller than another, three times more light; a surface four times smaller, four times more light on each square inch, and so on; the principle here being that the light increases or decreases on surfaces or areas in the inverse ratio of their sizes.

What is meant by this Law is clear enough, when thus stated; but when it is said that this increase and decrease of light is inversely *as the squares of the distance,* instead of saying that it is inversely as the size of the objects illuminated, it is not so clear that these words mean precisely the same thing. It is not so clear that the phrase "as the square of the distance" only means "as the size of the areas."

This expression, borrowed from Geometry, will be more fully explained in a future page, but thus much may be here mentioned: Every two areas or surfaces of different sizes that we find in Nature, are regarded in Geometry as two sections of an imaginary cone or pyramid, of which the greater area is as it were the base, the smaller area having its place as near the angle or apex as its size will admit of. Very little reflection will show us that their relative distance from the apex will depend upon their relative size—*i.e.*, will be determined by this relative size; but when the greater area, or supposed base of the pyramid, is four times as great as the other, the relative distance of this greater area from the apex is much less than four times that of the other. This base or greater area is then only *twice* as far from the apex as the smaller area, *two* being the square root of *four* (their relative size). In like manner, if, of the two surfaces which we compare, one has an area 900 times greater than the other, and we construct ideally their cone or pyramid, placing the smaller surface as near the apex as it will go, we find that the base or greater area is 30 times farther from the apex than the smaller one; 30 being the square root of 900; and so on, however great or little the difference between the two areas may be. And if, on the other hand, the base is four times farther from the apex (or centre) than the smaller area, we find that it is consequently (4 × 4) 16 times a greater area than the smaller is, because 16 is the square of 4; or if the base is 40 times farther from the apex than the smaller area, then the difference between the two areas is necessarily such that the base or greater area is (40 × 40) 1,600 times greater than the smaller. Or, again, if the base or greater of the two areas is 1,000 times farther from the apex (or centre) than the smaller area, the larger surface will thus be (1,000 × 1,000) a million times larger than the smaller; and, *therefore*, according to the "Law of the Inverse Squares," the light upon this *greater* area will be a million times *less* than it is upon each spot of the smaller; *i.e.*, in *inverse* ratio to the square of the distance. So that when we are given the relative size of any two surfaces in Nature,—the disc of Neptune for instance and that of Jupiter,—we can state the geometrical amount of their relative distance, *i.e.*, of their distance from the apex of their pyramid (for we thus suppose them to be sections of a cone or pyramid), and *vice versâ;* when we know their relative distance in the pyramid to which they thus geometrically belong, we know their relative size. This is the general law of all areas in Geometry; but of course it is only in the diagrams of that science that we ever see the cones and pyramids in question; while in Nature everywhere we have the surfaces or areas which require to be thus classified, and which can have their geometrical *distances* calculated by this law.

We see, then, that areas are (less or greater) as the squares of their distances from their geometrical apex; and instead of saying that Light is inversely as the areas illuminated, *i.e.*, is less when the areas are enlarged, we are taught to say that Light is inversely *as the square of their distance from their apex in Geometry;* an expression which has been still further "jumbled" by being abbreviated; for we now say simply that Light is *as the square of the distance;* most people fancying that the distance here meant is distance from the source of the Light; whereas what is meant is *distance from the geometrical apex belonging to the areas illuminated.*

The auditor or reader is not supposed to enter into such troublesome distinctions; but the lecturer or writer—at least the more enlightened one—is well aware that this is the purport of his words. He says that, as areas are directly as the squares of their distance from their apex, so the *illumination* of areas *must be* inversely as the squares of their distance from this apex. He well knows that this is the meaning of what he says, although his hearer or reader does not.

When one hears this, however, of two pictures or two tables, one naturally looks around for an apex or a pyramid, and wonders what such words can mean, or how an apex or angle can affect the illumination of tables on the floor or of pictures on the wall. Where is the apex,—where the geometrical distance,—of the two pictures before us, one of which has an area 4 times greater than the other, and, therefore, by the theory, 4 times less Light, although both close together on the wall and at the same distance from the window? Where is or what is this Apex,—this angle which makes one picture or one table bigger than the other?

The plain answer then is, as the reader now sees, that there is no such thing actually, but only geometrically, existing. It is not among the objects of nature that we are to look for this apex, but in diagrams; especially in diagrams of Cones and Pyramids. There we find it, and there only. This " square of the distance from a centre or apex," when applied to two such pictures or other areas, is merely a geometrical expression for their relative sizes,— a mere *façon de parler*, borrowed from Geometry and its Diagrams, in which the horizontal sections of a cone or pyramid, or the segments of unequal circles, are greater in proportion as they are farther from the apex or the centre, and as much greater as the square of their relative distance from the apex or centre is greater. This is the geometrical Law according to which we define the relative size of areas; but in Nature there are none of these Cones or Pyramids for areas to be sections of. The areas,—the sections,—are there; but we can only imagine their cones and pyramids, and relative distances. Nor do we ever, in Nature and in common life, measure areas (of pictures or tables for instance) in this way. We never say that they are in the same ratio as the squares of their distances. We merely say that one is so many times bigger than the other. To say that the size of pictures or of planets' discs increases as the square of their distance would be looked upon as a very far-fetched and frivolous introduction of our geometrical knowledge.

There can be no doubt that the term "distance" in this geometrical way of describing the size of areas, has misguided many teachers of Optics as well as students. When they hear that Light diminishes in proportion as "the square of the distance" becomes greater, instead of in proportion as the size of the surface becomes greater, they are apt to think that this is distance from the source of light, and that what is important, therefore, in this law is the distance from the source, not the size of the object. They naturally think of distance as diminishing Light in the usual way, when passing through a Medium; and, if we only suppose a luminous apex, why may not this distance, passed through, in the imaginary cone or pyramid, have the same effect, especially since it defines, and, as it were, produces, the larger area? There is also a theory current, as we shall see later, that, although every apex is not a luminous point, every luminous point in the Universe is a sort of apex to a sort of pyramid whose sections or areas, of course, always become larger and larger as their distance from the source increases. There can be no doubt, I think, that the associations thus connected with this term "distance" as what causes the diminution of Light in a medium, have helped to deceive many as to what this "sad jumble of words" means which is employed in this theory as a mode of describing the comparative size of areas; but there is no difficulty in seeing how this matter really stands. We easily see that the expression is only a formula, borrowed from Geometry, for saying that one area is so many times greater than another. On this point all scientific men and even all the more enlightened professors are agreed. All recognize the meaning of their theory to be that Light is greater and greater as the area illuminated is less and less; and that the Light becomes less and less as the area illuminated becomes greater and greater. On this all are agreed of whom we here require to take account. But in what way this diminution of Light results from an enlarged area, scientific men are not agreed among themselves; some suggesting one theory for it and some another; some even requiring no theory at all to help them; and, forgetting altogether that the diminution which is supposed to take place results really from enlarged area, they attribute it, at random, to the mere distance from the source, as in the case of a medium, without any enlargement whatever of the area illuminated. (*See* Appendix, No. 15.) But the more eminent of those who write so, have evidently only committed an oversight; although we must not forget that this oversight has often misguided inferior men. They admit, in other passages, that the area or surface on which the Light is thrown must be enlarged by distance or otherwise, to enable the diminution to take place according to their theory, and that if the distance, whatever it is, does not bring enlarged surface, it does not bring diminished Light.

Here then we see that, although in both modes of diminution we can, in a certain sense, speak of the diminution of Light *by distance*, yet in neither of these cases is it really *distance* itself that occasions the diminution, but a greater amount of Medium, *i.e.*, of Absorption, in the one case, and a greater amount of illuminated area or transverse space, in the other; both of which (both the greater absorption and the greater area) can be brought about either by an increased distance or by other means; the distance being either distance from the source when we are speaking of a medium; or distance in a pyramid, when we are not. The distance augments the medium in the one case; in the other case, it enlarges the section of the pyramid, *i.e.*, the space or area to be illuminated. On this account in both these cases the diminution of Light can be said, in a very large and loose way, to depend upon distance. This is easy;

enough to understand with regard to the lengthening of the medium, which of course diminishes the light; but less easy to understand with regard to the lengthening of the cone or pyramid. The band or sphere around a lamp or other source, becoming greater and greater as it is formed farther from the lamp, and therefore requiring (as they would tell us) a larger supply of light without receiving it, as well as the increasing sections of a pyramid, all equally exposed to exactly the same amount of Light, and receiving no more when large than when small, are matters a little different, it is true, from diminution in a medium, but still, upon a little reflection, quite as easy to comprehend the meaning of, although, of course, not so easy to assent to.

The important points here to attend to are (1) that, where all medium is supposed absent, it is the illuminated surface, thus enlarged, not the distance from the lamp, nor even from the apex of the pyramid, that is, in reality, supposed to diminish the Light, although the size of this surface depends upon this distance; and that where the area or space is not enlarged, the light in question does not expand, and is therefore not diminished, however great the change may be that is made in the distance; and (2) that, in the other case, it is the amount of the medium, or absorbing power, not its length alone, that diminishes the Light, although quite true that the amount depends, to some extent, upon the length of the medium, or distance to which it extends; for a dense medium with a short distance, or a thin medium with a long distance, would produce the same amount of Medium.

We all see easily how it is that Light is diminished when it passes through a Medium; and that the longer the Medium is, through which it passes, the more it is diminished, even if the Medium be invisible; but no one, as I have already remarked, pretends to know with much certainty what it is that diminishes light upon any given spot merely because another such spot previously dark is added to the illuminated one, or because (which is the same

thing) a larger space or surface is substituted for a smaller
one. Some of the causes conjectured for this diminution
by various theorists, I shall require to speak of in a
future page; but whatever this cause may be,—whatever
be the origin assigned to the decrease of Light supposed to
take place in connection with the enlargement of the space
upon which the whole force of the luminous body falls,—
this cause is considered to be strictly limited, in its operation,
to the enlarged space (thus requiring to be illuminated)
precisely as if this cause consisted in some infinitely expansive nature in the light itself, arrested only by want of room
to expand in. The same amount or degree of Light, which
illuminates a small space, becomes in this way stretched
and attenuated as it were, by this curious theory, and thus
diminished (not relatively, as one might imagine, but absolutely diminished) UPON EACH SPOT, just in proportion as that
space is enlarged; for without this proportionate enlargement no diminution can result; and the Light, we are told,
becomes less and less in such a case, exactly in proportion
as the illuminated space becomes greater and greater, or to
express it in the favourite way, as the square of the distance becomes greater and greater. We can, of course,
easily understand how this should happen if the source itself were subdivided and distributed or scattered; but in
the case we are speaking of, the Unity or centrality of the
source is the very thing to be dealt with, and the equal expansion or diffusion of the Light over the enlarged space,
is supposed, in the theory, to be combined with this Unity
of the source.

Enough has been now said to enable the reader to comprehend sufficiently what is meant by saying that Light is
diminished by Absorption or Medium, and what is meant
when we say that it is diminished by Expansion, or Enlarged Space; and he must bear carefully in mind that
there are thus supposed, by the ablest writers in our
country, to be two very different processes of diminution
always resulting to all Light from Distance, and both,

where this distance is concerned, always going on at the same time, and almost always, both of them, very considerable; these being (1) the diminution resulting from the Medium, and (2) the diminution resulting from *mere addition* to the area or quantity of space illuminated; each kind of diminution, moreover, it will be seen, has its own special Law, each Law being entirely different from the other, but both Laws taking effect in every case of Light diminished upon some principle of distance. The Law in the case of the Medium is called a "Geometrical Progression," whereas the Law of diminution in the other case is commonly called the "Law of the Inverse Squares." It is necessary that the reader who seeks to understand this subject should clearly understand these two Laws; and they are in themselves so easy that no one need turn away from them, unless indeed they were to appear before him in the cabalistic terms of the Profession; which, in these pages, they shall not.

THIRD SECTION.

EXPLANATION OF THE LAW FOR THAT DECREASE OF LIGHT WHICH RESULTS FROM A MEDIUM.

In the case of the Medium the unscientific or popular notion is that, in a uniform medium, the diminution increases exactly as the distance does (but more or less rapidly according to the density or thickness of the transparent substance which we call the Medium)—that, at twice the distance there is twice the diminution; at three times the distance, three times the diminution;—that when one distance is four times greater than another, the diminution in the Light is also four times greater; and when thirty times greater, the diminution also is thirty times greater. According to this rule, each equal distance, in the uniform medium, produces the same effect. Each equal length of medium abstracts the same amount of light from the whole original amount. If, for instance, the reduction, in the first equal length or distance, is one-half of the whole light, then the reduction in the second distance will also be one-half of the whole original light ; and the whole Light is thus finished, there being no more to pass into the third distance. Or if the first reduction is $\frac{1}{1000}$ of the whole light, and the medium is long enough to have 1000 such reductions, then after the last of them, the whole light is absorbed. This is the common and natural or "unscientific" view of this matter.

Many of the scientific, however, look upon this as an error. If it is considered more correct to say so, I am quite willing to say that they all do so. Certainly all the Professors do so. They hold that the reduction which takes place is of a different amount in each of these equal distances, although the Medium is (as here supposed and as is indispensable for the calculation) perfectly uniform. They hold, moreover, that the Light can never become completely extinct by the action of the Medium. Their Law is that if the reduction, in the first distance, is one-half of the whole Light, then the reduction in the second distance (although the distance is equal and the Medium the same throughout) will only be half as much as in the first distance, leaving still a quarter of the original light; and that, in the third distance, the reduction will be half this quarter, or one-eighth of the whole original light; and so on, *ad infinitum*, each succeeding distance absorbing exactly one-half of the light immediately preceding it, whether that which preceded it were much or little; never any final extinction possible; and no two distances (*i.e.* no two equal amounts of medium) producing the same effect,—no two producing the same amount of diminution; and the same principle of diminution is observed when the first reduction is only one-thousandth of the whole light. The second reduction would not be one-thousandth of the whole light, but only $\frac{1}{1000}$ of what was left after the first; and so on.

This is called a Geometrical Progression, and a very curious Progression the reader will probably be inclined to think it. In it each distance is supposed to absorb the same *fraction* or proportion of the Light immediately preceding it as was absorbed by the preceding distance, but not, by any means, or ever, the same *quantity* of Light; and in it the calculation is so arranged that no medium, however dense or extensive, can ever absorb the whole light.

The unaccountable misapprehension of Nature involved in this Law and Theory for the diminution of Light will be pointed out more fully in the Fourth Part of this Treatise. All that is here important to be taken account of, beyond what I have just mentioned, is that this is a Law and Theory of *Reductions*, not, as in the other case to be now explained, a Law and Theory of the *Quantities that undergo these Reductions*, and which exist before or after the Reductions have taken place. This distinction between the two Laws, if carefully preserved and attended to, will be found to facilitate the study of this question. Quantities and the Reductions to which they are subject are entirely different things, and have laws entirely different.

FOURTH SECTION.

EXPLANATION OF THE LAW CALLED THE LAW OF THE INVERSE SQUARES; VIZ., OF THE LAW FOR THAT DECREASE OF LIGHT WHICH IS SUPPOSED TO RESULT FROM THE ADDITIONAL SPACE ILLUMINATED.

General Meaning of the Law.—This law, according to which Light is supposed to diminish as the square of the distance increases, is to the effect that, apart from all medium and its action, the same sum total of Light is more and stronger upon each spot of a smaller object or area than upon each spot of a larger one; (or, which is, of course, the same thing, less and feebler upon each spot of the larger object or area than upon each spot of the smaller one), and is so precisely in the proportion of these areas. Thus the Light from a unit-source, is, upon each spot of each area or object, according to this theory, in inverse ratio to the size of the objects illuminated,—great when they are little, and little when they are great. That is the whole import of this celebrated law.

The language adopted, however, to express it, contains a little entanglement, and leads to a great deal of confusion, being justly called, by one of our most distinguished Physicists, "a sad jumble of words."

When the reader hears that "Light is inversely as the square of the distance," or that "it diminishes as that square increases," he naturally asks, What distance? and when he is informed that this distance is only a geometrical measure of the smaller or larger areas illuminated, he also asks quite naturally, What can this distance or its square have to do with the size of the objects illuminated?

All this matter is more fully explained further on; but it is necessary that some idea be here given of what is meant. The reader knows probably that the square of any number is that number multiplied by itself. Thus 4 is the square of 2; 9 is the square of 3; 144 is the square of 12; and 900 is the square of 30; and it is found that this is the proportion which exists in geometry between the sections of a pyramid or cone and their distances from the apex, corner, or angle by which the relative sizes of the sections are determined.

It is obvious that if we cut an upright cone or pyramid horizontally at different distances, the sections are greater and greater the lower down we make the cuttings, i.e., the farther they are made from the apex; and it can be proved that these sections or areas are larger or smaller than one another,—not in the same ratio as their relative distance from the top, as one might at first suppose, but in the same ratio as the *square* of this distance. Thus, if one section is twice as far from the apex as another, it is also (2×2) 4 times larger than that other, instead of being only twice as large; if it is 3 times farther from the apex, it is (3 × 3) 9 times larger; if 30 times farther, the section or area is not 30 times, as one might expect it to be, but (30 × 30) 900 times larger, and so on.

We thus see that, in a cone or pyramid, the relative proportion of any two sections or areas is always exactly *as the square of the distance*, *i.e.*, of their relative distance from the apex, the size of whose angle gives them their relative sizes, and that therefore the square of their distance can be employed as their measure, and even as a synonym for their relative size. We say therefore that these areas or sizes are as the square of the distance; and since from any two objects in nature, of different sizes, we can, in geometry, suppose their pyramid constructed; we see that we can say the same of all sizes and of all objects, viz., that they are as this square of the distance from the corner which fixes their relative distances.

Since this is true then of all areas or sizes of objects around us in nature, as well as of the sections of a pyramid or cone,—since all the areas or sizes of objects are thus as the square of the distance, *i.e.*, of their distance in the geometrical diagram,—of their distance from a supposed apex or determining corner,—we, for this reason, say that the objects (meaning always their areas or sizes) increase or decrease in that ratio; so that anything which decreases in proportion as these areas increase, and increases in proportion as these areas decrease, can be said to increase or decrease inversely as the square of the distance, or, simply, to *be* inversely as that square. Thus we can say, for instance, that any two surfaces of different sizes (walls, fields, pictures, &c.) are as this square of the distance, *i.e.*, are greater or less in exactly the same proportion as this square; and that, if the two surfaces are boards and the same amount of paint is to be spread over each, this paint will be (in thickness upon them) *inversely* as the square of the distance, in other words, thicker upon the smaller board

and thinner upon the larger one according to the difference
of their sizes, or, in other words, and geometrically speaking, according to the square of their distance. Thus also
when, in this theory, it is wished to say that a given amount
of Light, if spread over small areas, is "thicker" and
stronger than if spread over areas that are larger, the
theorists say that the Light increases and decreases inversely
as the square of the distance; meaning thereby inversely
as the size of the areas or objects illuminated, being
stronger (thicker) where these are small, and feebler
(thinner) where these are large. In other words : Since in
geometry and in geometrical language, every two objects or
surfaces of different sizes, that we see around us, belong to
what in that science is called a pyramid or cone, and since
all the objects around us, and their areas, are in this geometrical relation to one another, viz., as the square of their
relative distances from the apex of their own special hypothetical, or imaginary, pyramid, for these reasons, the
illumination of these bodies, instead of being said, in plain
language, to be inversely as the objects or areas upon which
the same quantity of Light falls and spreads, is, in this
obscure law, said to be inversely as the squares of their
relative distance from the apex of their supposed pyramid ;
a statement which is, of course, only the same thing in other
less familiar words,—but in other words so obviously calculated to deceive us, as to make us naturally suspect that
those who originally thus expressed it, were themselves led
astray by the very expressions they employed.

Of " Spreading " and " Distance " in this theory.—There are
two terms in the foregoing exposition of the Law which
require to have the reader's attention especially drawn to
them, as they are apt to involve equivocation ; the term
" spreading " and the term " distance." Neither term here
means what it seems to mean. The spreading meant in this
theory is never the spreading of Light from the luminary
to the object illuminated, but only that which is supposed
to take place from one side of the illuminated object to the

other side of it; and the term " distance " here never means
the distance between the luminary and the object illuminated,
as we may see by the fact that this distance could never
make the object larger or smaller as distance in the pyramid
does, whereas it is this size of the object, that, in this
theory, increases or diminishes the Light.   The distance
spoken of in this Law and Theory never means the distance
between a picture, for instance, and the window or other
source of Light.   The distance here meant can only exist·
when there are two pictures of different sizes, wherever
they may happen to be placed, and has no reference whatever to the window or to any other source of the light falling
upon them, nor to any distance to be found in the room.
It denotes merely *the geometrical position* belonging to the
sizes or areas of these pictures, *viz., the distance which is
greater or less according as the two pictures differ in size from
one another.*

The principle or measure called " the square of the distance " is here and everywhere only used to define or
describe the comparative sizes of the objects around us, in
nature (whose sizes are all classed as the sections of a
pyramid), and has no reference of any kind to their distance
from a source of light.   The theorists, when they use this
expression, merely mean that the intensity diminishes when
the object is greater on which the light falls and over which
it spreads.   That is all that the more intelligent and enlightened writers mean.   " The square of the distance" has
no reference whatever for such men, except to the greater
or smaller objects around us, upon which the light is supposed to spread.   It has nothing whatever to do with their
distance from the window, lamp, sun, or other luminary, as
has been absurdly supposed by so many professors
apparently unacquainted with the import of the geometrical
formula; the term "distance" in which formula merely
affords, as I have said, a measure of the area of such objects,
and is, in this law, only used as such.   But I need not dwell
upon this point.   The least reflection suffices to show that

no distance from a source of light can make one area larger or smaller than another, this difference of *size* being that *alone* which, in this theory, increases or diminishes the illumination. The square of the relative distance from the apex in a pyramid defines exactly, for instance, the relative size of two pictures; but the relative size of no two pictures could ever be discovered or defined by our knowing their relative distance from a lamp, a window, or the sun.

The sense in which Light is supposed to " spread," in this theory, does not present so much chance of misconception as the sense in which the word " distance" is employed; yet it has its vagueness also, and on this subject no vagueness can be allowed. The spreading of Light is always understood, by these theorists, to be, as above remarked, right and left, or up and down, upon the surface illuminated; but not from the source outwards to that surface; for the spreading in question is not required in that direction, that being the natural path of the rays, which proceed along it, undiminished in force or number, to the confines of the system. It is thus held that Light spreads in two dimensions of space, as it were—not in the third. The law we are now speaking of has no reference, we see, to a Medium, nor to the action of a Medium. This must be well borne in mind. The spreading here meant, therefore, is not that result of a medium with which we are so familiar, and which enables Light not only to extend, but even to go round a corner. The spreading of Light, in this theory, is supposed to take place independently of all medium. The theory supposes the light, from any given central source, to be the same, in amount and degree, upon surfaces of all sizes, equally exposed to it, however much they differ from one another in size, and however nearer some may be from the source than others. For instance, the light is, by this law and theory, exactly the same at one mile from a lamp as it is at three feet from it; and the amount which falls upon an area a mile square, all equally exposed to the lamp, is the same as that which falls upon a square

yard. The law then states that as this same amount of
light is in the one case spread out over a much larger
space than in the other, the light upon this larger space is
thereby diluted, and rendered feebler in proportion as the
one space is more extensive than the other, or, as the
favourite expression runs, has a greater square of distance.
Since the illumination is thus held to be inversely as the
surfaces are great or small,—*i.e.*, as they have a greater or
less square of distance,—the law states that it is inversely as this size or square of distance;—a square of
distance, however, which the reader now sees has nothing
whatever to do with the distance of the object from the
window or other source of light.   The law as well as the
theory is, that Light loses in intensity when the surface it
has to cover, or be spread over, is enlarged (*i.e.*, has a
greater square of distance), however near the source this
larger surface may be, and gains in intensity when the surface
it has to cover, or be spread over, is reduced in size (*i.e.*, has
a less square of distance), however distant from the source
this reduced surface may be placed ; the illumination depending, as just said, not on the distance or proximity of
the source, nor on any spreading of the light from the
source *towards* the illuminated object, but wholly upon the
size of the areas illuminated, or, in geometrical language
(rarely used by geometricians themselves, but always here
by physicists), on the square of their relative *distance* from
the apex of their pyramid.

Such is the Spreading here intended, and such the Distance whose square is mentioned in the Law.

*Difference in point of Distance between this Law and the Law
for Diminution in a Medium.*—From what has now been
said, the general reader (for whom mainly I write) will have
no difficulty in seeing that the Law and Theory now under
consideration—the Law and Theory of Diminution, in the
second or hypothetical form alluded to in the Second Section
of this Part, and which may correctly be called the
" spatial " diminution, or diminution by the size alone or

space of the objects illuminated—differs, in this particular
point of Distance, *toto cælo*, from the diminution which
results from the Medium. This difference, of which it is
important that we here take strict account, is that, whereas
in a medium the length of the medium, or distance from
the source, always does as much as the density itself to
diminish the light, in this other mode of diminution by size
of surface (or, as it can be also expressed, by square of
distance), it does nothing. In this second theory, this so-
called "distance from the source" has not the smallest
effect in producing the diminution which is supposed to
take place. The distance spoken of in a medium is distance
from the source of Light; the distance spoken of in the
"square of the distance" is not. This latter distance is
only distance from the source of size. The term, in this
latter case, is only used as a measure and definition of the
areas or surfaces illuminated. It has nothing whatever to
do with the light or its diminution, or with the distance of
the luminary. We must therefore be very careful to observe
that when we speak of the "square of the distance," as is
so often done in connection with this second kind of diminution, we *always* mean only the size of bodies, this being
also called, in geometry, the square of the distance. The
expression never means their distance from the source
which lights them. That the distance from the luminary
does nothing, in this theory, to diminish light, all the more
enlightened and intelligent of our writers and lecturers,
when closely questioned, fully recognize. They all hold
that, in each case of this new theory, the diminution is
effected *only* by the enlarged area, where such can be
found. They do not, however, enter into any explanation,
either as to why there should be this necessary relation
between Light and Space, or as to what expansive or
elastic nature this is which they thus ascribe to Light, and
which renders it possible for it to preserve this relation by
spreading out and becoming attenuated, nor why, in fact,
there should be any need of the "spreading" when the

source is a central and unspread one, to which each portion of the whole surface is equally exposed. They evidently discern no physical impossibility in what they teach about this spreading, nor do they see that there is no blank to be dealt with,—no occasion whatever for the hypothesis of a spreading Light. They merely reiterate, as they do *ad nauseam* (and apparently without knowing the meaning of what they say), the geometrical principle respecting the sections of a pyramid and their distance from its apex and their size depending upon that distance,—all which, it will be seen, has nothing whatever to do with their theory, being, in fact, as they acknowledge, the very inverse or opposite of their theory, yet what they in the most frivolous and superficial manner combine with it and imitate in it. These errors will be fully pointed out in PART II. of the Treatise. I am here engaged only with the Exposition of the Law and Theory for the less scientific reader,—the first complete exposition, I am inclined to think, that any one has ever thought it desirable to give of them; and I am still far from having completed it.

The " Square of the Distance" in Geometry explained.—As, however, the geometrical principle in question has been so productive of confusion in the hands of our ungeometrical physicists, as well as in those of our geometricians but slightly acquainted with physical science, it is well that the general reader should see exactly what it amounts to, and how innocent it is of all the absurdity with which it has been so unsparingly interwoven. It is proper, therefore, to explain here more fully this matter of the concentric spheres and of the pyramid,—the basis of the geometrical principle in question, respecting the relative sizes of objects, —a clear knowledge of which will go far towards counter-acting the extensive misapprehensions current in this de-partment of Physical Optics, and rendering the exposition easier in which we are engaged.

Explained by the Concentric Spheres.—If we suppose several concentric spheres at equal distances from each other ; in other words, several hollow globes, one outside the other, and so placed that the centre of the inner globe serves as the centre for all, then it is found, in Geometry, that these spheres are in the same proportion to one another as the squares of their distances from this common centre are ; *i.e.*, it is found that when one sphere is *twice* as far from the centre as another, it is not then twice as large, but (2 × 2) 4 times as large as that other ; when 3 times as far from the centre, it is not 3 times as large, but 9 times as large ; when 4 times as far, it is 16 times as large, and when 30 times as far from the centre as the first sphere, it is (30 × 30) 900 times as large as that first sphere. The distance, in this case is called, in Geometry, the radius of the sphere, because it extends from its centre to its circumference ; and the proportion now indicated is known as the geometrical law of areas.

Explained by the Cone or Pyramid.—Or, to make the same statement respecting the sections of a cone or pyramid :— If we form a triangle with two equal sides, *i.e.*, an isosceles triangle, the two equal sides of which represent the sides of a cone or pyramid, a line drawn from the apex perpendicular to the base marks the distance of the base from the apex.

Now, if we divide that perpendicular line (or distance) into two equal parts, and draw a line, through the point of bisection, across the triangle or pyramid and parallel to the base, we shall thus have, inside the triangle, a second base, smaller than the first. The law here is that as the larger base or area is *twice* as far from the apex or vertex as the smaller one is, it is therefore (2 × 2) 4 times greater than the smaller one.

Again, instead of dividing the whole perpendicular or distance into two equal parts, divide it into three such parts, and draw the lines across, as before, parallel to the larger base. We then have three bases, or areas, sections of a cone or pyramid, the smallest base being 4 times less than the second, and 9 times less than the third, or largest base.

And again, if, instead of dividing the perpendicular into 3 equal parts, we divide it into 4 such parts, we then have 4 bases; and the smallest is now, not only, as before, 4 times less than the second, and 9 times less than the third, but also (4 × 4) 16 times less than the fourth.

Or, instead of dividing the perpendicular into 4 equal parts, let us make a larger diagram, and divide the perpendicular into 30 equal parts. Then as the smallest base or area is 30 times nearer to the apex of the cone or pyramid than the largest base or area is, it is also, and for that reason, (30 × 30) 900 times smaller than the largest; or, which is, of course, the same thing, the larger base or area is 900 times greater than the smaller is, that number being (30 × 30) the square of its greater distance; and so on, of all other areas, to any extent that we choose to subdivide or lengthen the perpendicular or distance. And this is what we mean when we say that these areas or sections of the cone or pyramid, in geometrical diagrams, are always, not only greater and greater as they are farther from their apex, but are so in the same ratio as the square of their relative distance from it.

The Square of the Distance applied to Natural Objects.— Here, however, the important distinction already pointed out must be carefully preserved. The proportion of distance, now described, is that by which, in a pyramid or other geometrical diagram, the area is made of a certain size, and is known to have that proportion to another area; but in nature and common life, this is never known so. In nature this size can only be known by being measured. There is in nature and in common life no such distance, and no apex to give sizes to objects, or to be a measure of them. We must not imagine, as so many do, that the geometrical proportion of objects to their distance from an apex or a centre exists anywhere, except in the geometrical diagram. It is clear that no other areas, objects, or surfaces,—none of the surfaces of bodies in nature,—can be spoken of in this way. There is, I repeat, no apex, corner, or centre to refer them to. In geometrical diagrams one area is 4 times greater than another, because it is twice as far from the apex or the centre; but the area of a table or of a room is 4 times greater than that of another table or room, without any possible reference to a common centre or a common apex; and one planet is larger than another,—the Earth, for instance, larger than Mars,—without any such reference. The misconception I now advert to is an error here made by all the Professors. Geometry teaches that the imaginary areas (sections) of a pyramid or of a cone are thus proportioned to their relative distance from an actual common apex in the diagram; but this does not mean that other areas are so. This does not mean that the other areas or surfaces, not in diagrams but around us in the world, in nature or in art,—the walls of our houses or the discs of the planets,—are thus proportioned to their distance from some actual common apex,—nor that this sort of thing ever happens except in geometrical diagrams or in other artificial arrangements. All the bodies in nature are without this proportion to distance, having their exact size quite independent of their position, and determinable by measurement alone.

It is clear then that our business respecting this square of distance is with areas or surfaces only, and not at all with their actual distances either from each other or from anything else; but we must now attend to what we mean when we say that one area is a certain number of times greater or less than another, as it is this which is supposed to augment or to decrease the degree of illumination according to the law of the Inverse Squares.

The surface of the second concentric sphere, or of the second section of the pyramid, being thus 4 times more extensive than that of the first, it has 4 square inches or square feet, for each square inch or square foot of the first; and the surface of the third sphere or section being 9 times greater in extent than the surface of the first, it has 9 square inches where the first sphere or section has but one; and the thirtieth sphere or section, being 900 times greater than the first, has 900 square inches or square feet for every square inch or square foot of the first; so that there is that much more space to be covered in the outer sphere than in the inner one, or in the larger section of the pyramid than in the smaller one, by heat, air, or other element seeking its own level (as water does), or by the paint or tar that we might employ to spread over it.

A given amount of heat would thus be lower upon each inch or foot, and of tar or paint, thinner and more diluted, according to the distance from the centre or apex, and according to the square of that distance, but wholly irrespective of where this heat or tar or paint may come from; and at 30 times a greater distance from the apex or the centre, the same amount of tar would be 900 times more diluted or thinner, at each point, on the surface tarred, than at the first distance. It is evident that where with two equal quantities of tar, we have two areas to be covered all over, to the same thickness upon each spot of each, these areas must be equal; and that, in order to bring about the sort of diminution in the tar which is here described, one of the two areas must be made larger than the other. No such diminution or thinning out can take place without this enlargement of one area. It matters not how these areas are placed with reference to one another, nor what their sizes are, nor what the difference between them is, provided there is *some* difference, and provided the whole quantity of tar is precisely the same for each area. These are the essential conditions. The true law then here is, that the quantity of tar being exactly the same upon the *whole* of each area, its amount *upon each spot* of the larger area is less than that *upon each spot* of the smaller, in proportion as the larger area is greater than the smaller one. This is the Law of the Inverse Squares. In this strange phraseology, the tar is said to be inversely as the square of the distance; *i.e.*, inversely as the areas tarred;—thin when the area is large, and thick when the area is small. It is

entirely contrary to this " Law of the Inverse Squares," as
it is called, to suppose that, the two areas being of the
same size, the tar, *on each spot*, of either could in such a
case be diminished, at whatever distance they might be
from one another, or from anything else ;—also entirely contrary to this law to suppose that, when the areas are or
become different, the amount of tar is different also. The
Law fully recognizes the fact of nature that the degree or
amount allotted to each area is completely and permanently
the same degree or amount for each area, however enormously they may differ from each other in size. Yet the
scientific expression ordinarily given to the proportions of
this fact is, that the tar diminishes inversely as the square
of the distance ; *i.e.*, becomes less and less in amount as the
area becomes greater and greater ;—no very remarkable
instance this, it will be admitted, of " scientific" accuracy
or of " scientific" statement. The amount on each area
remains by this theory exactly the same. The obvious fact
which here requires to be expressed is not that the tar is
less when the area is greater ; for that would simply be
false (the hypothesis being, as I say, that the whole amount
remains the same upon each area) ; the fact which here requires to be expressed, is that the area being greater, there is
therefore less of the tar *upon each spot*. The expression
" upon each spot," or upon each " unity of surface," as it is
technically expressed, is the whole secret of the statements.
This sort of reduction *at each point* of the larger surface,
while on this surface, *as a whole*, there is no reduction at
all, is what, when plainly stated, the most uneducated
workman can easily understand. It is, moreover, what
nobody denies.

Now it is this law of dilutable or dilatable elements upon
areas of different dimensions, which Physicists delight to
call *the law of the Inverse Squares*, and which they have
hitherto transferred to Light, although Light can neither be
diluted nor dilated ; some of them even seeking to apply it
in this way without the necessary enlargement of tho

space. (*See* Appendix, No. 15.) The ball of tar or the pot of paint, or the bit of gold to be beaten out, or the gallon of self-diffusing water, or the degree of self-diffusing heat, being entirely cut off from the quarter whence it comes, has to be spread out (either with brush or hammer, or merely by being left to its own self-diffusing nature), over a greater extent at each enlargement of the area, or, which, in geometry, is the same thing, at each remove from the centre or the apex. Without this enlargement of the area, the water, paint, or tar cannot be diminished in quantity upon each spot. This is a most important point to be attended to; for, as observed above, some writers when applying this theory to light—men too of great eminence—have written as if they fancied that the enlargement of the area was not necessary to bring about this sort of diminution provided there were an increased distance from the source of light, as in a medium; whereas this enlargement of the surface is the *sine quâ non* of the common fact with regard to water, paint, or tar, and must therefore be so of the theory which applies exactly the same principle to Light, teaching that Light also diminishes in this way. These writers entirely misapprehend the theory when they suppose (*ibid.*) that the enlargement of the area illuminated has nothing to do with the diminution of the Light, and when they represent that what alone is wanted is *distance* from the luminary. It is quite otherwise. What alone is wanted by the theory is the enlarged area. The distance from the luminary has nothing whatever to do with the diminution. The sole distance in question is one unconnected with the luminary, while the sole object here of this distance is to describe geometrically the greater extent of area, and to indicate the proportion we wish to speak of in the areas. If the distance mentioned by the Physicist does not give an enlarged object, it does not comply with the geometrical conditions of his theory; and it cannot give an enlarged object if what he speaks of under the term "distance" is distance from a luminary.

Error of supposing that Light is diminished in this theory by distance from the Source of Light.—This error of supposing distance from a luminary the cause of diminished Light (where no medium is taken account of), and which, at least in their language, is common to all these writers and lecturers, is well illustrated in the passage just referred to (Appendix, No. 15), where only distance from a source of light, and no enlargement whatever of the area, is considered necessary for the diminution of Light upon each spot of the area. The same writer however, it will be seen, in the same Extracts, recognizes, or, at least, illustrates his own error, in the instance of the boards employed as screens, and also in that of the two boxes with a taper in the centre of each. He there points out distinctly that, in the theory, the Light is the same in amount on the two illuminated areas compared, in each case, however different in size these areas may be; and that, in order to produce the diminution of Light upon each spot, as taught in the theory, these areas, in each case (both in the screens and in the boxes), must be of different magnitudes. (*See* also Appendix, Nos. 2 and 3.) If the Earth were twice as far as it is from the apex of its supposed pyramid or from the centre of its supposed sphere, and had, as in Geometry would on that account be indispensable, a disc four times as large as it has in its present position, its solar light would be reduced, by the received Expansion-theory, to one-fourth of what it is; but if the area is not enlarged, how is even the reduction mentioned in this preposterous theory to take place? At that distance, either the Earth's disc would have to expand to an area four times larger than it is, in order to have this diminished Light upon each spot, or, if it remains of the same size in its new position, it becomes necessary to compare it with a disc four times smaller, at its former distance from the sun, in order to bring it within the theory, and to give it the geometrical proportions of the law.

What changes here with the distance, according to the

law and theory now being described, is not the amount of
light allotted to each area (for that, by the theory, is always
the same at every distance), but the area itself, or extent
over which the same light is supposed to spread.   No one
pretends that the amount of light changes.   No one pretends that there is more on the small area than on the large
one, nor near the source than at a distance from it.   They
only say that it has to cover more space, and so thins out,
or becomes "diluted."   If, at the greater distance, there is
not the greater area, how could there be, I again ask, even
the diminution which the theory professes to find?   It is
the enlarged area, then, not the distance, which is required
for the diminution of the Light according to this theory.
The introduction of the term "distance" is only required
for the mathematical expression "square of distance," and
the air of mathematical certainty thence derived which has
allured and misguided so many Lecturers.   Nothing however can be clearer than this part of the theory, although
some of the theorists themselves, in their writings and
lectures, seem to have been considerably puzzled by it.
The real difficulty of the theory is, when we have the enlargement of the area, how are we to conceive of that as
diminishing the Light?   For, unlike air, or water, Light,
all admit, neither seeks an equilibrium, nor can, unless in a
medium, be made to leave the straight line at all, or to
turn corners; as is well seen and fully recognized in the
umbra and penumbra of the eclipse.

This point respecting the "spreading" or "expanding"
of Light, without a medium, or even with one, will be found
fully discussed in PART II.   It is sufficient here to have
adverted to it in distinct terms as the most essential part
of the theory now described, nay, almost as the whole of
that theory; for, as we have seen, the squares and distances, put so much forward by the theorists, as part of it,
have really nothing whatever to do with it, being merely
the geometrical way of describing the larger or the smaller
area.   But although the spreading by expansion, with the

consequent "dilution," is the whole theory, nevertheless
the professors of Physics among us omit all allusion to this
point as attended with the least difficulty, and in their specu-
lations respecting Light, treat it as a clear and recognized
fact that Light, quite as naturally without the action of a
medium as with it, diffuses itself and goes round corners,
and is subdivided independently of its source, *i.e.*, without
the aid of any diffusion, spreading, or subdivision in the
source. The effect, which the whole power of the unspread and undivided source can produce on each square
inch fully and equally exposed to its action in the smaller
sphere or surface cannot, it is supposed, be produced on
each square inch similarly exposed to the same power, in
the larger sphere, on account of the *spreading* that would,
they say, necessarily result from the larger sphere and of
the *thinning out* that would result from the spreading. This
is the theory of the Inverse Squares ; the larger the surface
and the less the Light upon each square inch of it. In
other words : To diminish the Light that there is on any
area, we need only enlarge the area. The Light then, of
itself, stretches, spreads, and so becomes thinner. As long
as the area remains of the same size, no matter at what
distance from the source we place it, we can have in this
theory no diminution of the Light, because we can thus
have no stretching or spreading.

*How Light is diminished through Enlargement of the Area,
shown by the Concentric Spheres and the Screens.*—The example
of the spheres given above (pp. 24, 25) to explain the geometrical proportion between areas and distances, may be here
profitably repeated to explain the diminution of Light supposed, in this theory, to ensue upon the enlargement of the
illuminated area. Let us suppose any number of these
concentric spheres of space with a lamp or the sun in their
common centre, and each at the same distance from the one
preceding it as the innermost is from the lamp or the sun.
The professors here admit, however vaguely they express
themselves, that the *whole* Light of the sun or lamp is upon

each of these spheres—upon the thirtieth of them, if we suppose thirty, as much as upon the one next the lamp or sun; and therefore precisely *the same amount* of light on each of them, however much they may differ in size; *i.e.*, precisely the same amount of light on areas of all sizes that are equally exposed to the same source,—on the area (30 × 30) 900 times greater than the area next the luminary, as well as on this smallest area itself. It is important that the reader should well understand all this. None of these theorists or professors imagine that there is less light on a small area than on a large one when both are equally exposed to the same source; nor less on a large one than on a small one, whatever the geometrical description of such an area may be; *i.e.*, however many times the square of the distance, in the case of the one area, may exceed that in the case of the other. They all admit the following propositions to be part of their theory although they do not all say so as distinctly as if they were anxious their words should so be understood

1. That (omitting of course, as always in this theory, all effect of Medium), every area or surface has upon it the *whole* light that the source can give to which it is fully exposed; and this, irrespective of *distance*, and irrespective of *size;* the one which is thirty times more distant from the source, as well as the one next the source, and the very small area as well as the very large one.

2. That (even under the influence of a Medium) there is precisely *the same amount* or degree of Light upon areas or surfaces *of all sizes*,—upon a very large area and upon a very small one,—when they are all equally exposed to the same source; as, in this example of the concentric spheres, there is precisely the same amount of Light upon the smallest area as there is upon the area 900 times greater than the smallest, neither more nor less on either; and that, in the absence of all medium, this perfect equality of light for all areas subsists also *at all distances* from the source, the most distant areas having, by the theory and by this example

of the concentric spheres, precisely *the same amount* of light upon the whole of each area as those nearest to the source.

3. That it is the enlarged area, not its distance from the sun, which, in the case of these concentric spheres, is supposed to diminish the Light; as may be seen from the fact that the diminution upon the largest area is said to be 900 times greater than upon the smallest, this being the proportion given between the two areas in that case.

4. That this diminution of light upon the enlarged area results from the *expansion*, or distribution, over a larger surface, of the same amount or degree of Light as falls more condensed upon the smaller surface, which is equally exposed to the same source, the light being thereby proportionably diluted and thinned out; and that this effect results without any expansion or distribution of the source.

The professors consider that since, in the present illustration, the whole power of the source falls with a certain effect upon each square inch in the smallest of our thirty spheres, it must necessarily fall with 900 times less effect upon each square inch of the sphere 900 times larger; either because, without thus economizing itself, it might not be able to supply this action of its whole nature on each of the square inches thus increased in number; or because the source, although in reality a unit, undivided and unspread, is nevertheless *to be thought of*, for the sake of the theory, as divided, spread out, and distributed, among the different square inches of each sphere, and each division of the source as giving its own separate and independent part or amount of the Light; which supposed part, division, or portion of the source has always of course to become more and more minute upon every fresh enlargement of the space or surface which has to be illuminated with only the one given amount of Light. In this manner they consider that the whole united source which illuminates with its whole strength each square inch in the first of our concentric spheres, can only effect the illumination of the sphere which is four times greater by *dividing* its energy so as to be four

times less upon each separate square inch of the greater
sphere, although this source is itself entirely undivided,
unspread, and unreduced; and that in like manner, when, at
the third distance, it has to operate over a surface nine
times greater than that over which it had to operate in the
first sphere, it then requires to divide its energy or power
into parts as much smaller as the square inches are more
numerous; in other words, it can only exist then, in one-
ninth of its whole force, on each square inch, just as if the
source had been divided and distributed for the purpose,
although the whole of this source remains together and
acts together upon each point exposed to the whole of it.
In like manner, as already said, the largest of our thirty
concentric spheres,—the sphere 900 times greater than the
smallest,—has its light, by this theory, *not* 900 times less
than the smallest (as the careless express it), because it
has, all acknowledge, precisely the same amount of light
upon it as the smallest has, and is equally exposed to the
whole source, *but* 900 times more diluted and thinned out,
on account of the larger space and the consequent spreading,—900 times less than the smallest, *upon each spot*, or *at
each point* of its surface.

Another and in some respects, better mode of illustrating
this theory for the diminution of light without a me lium, is
that of substituting screens for the concentric spheres, as
most Lecturers and writers do. (*See* Appendix, Nos. 2, 3, 15,
17, and 20). Some of the peculiarities of the theory, too liable
to be overlooked, become thus more manifest. (1.) In the
hypothetical concentric spheres, for instance, we have no atmosphere to deal with; whereas, with the screens, we have
the reduction of Light resulting from our dense atmosphere,
and going on at the same time as that supposed to result
from enlarged space, *yet producing no difference whatever in
the result;* a combination of agencies bringing considerable
confusion to the aid of the theorists, and of which they
always avail themselves when they are allowed to do so.
(2.) The screens show, more clearly than perhaps the

spheres do, that it is the greater area or greater trans-verse space alone, irrespective of form (whether plane or spherical, round or square), which by this theory occasions the feebler light ; that where the area is not enlarged there can be none of the diminution in question, and that according to this theory the square yard of surface has less light on each square foot of it, than a square foot of surface alone has on it, when, both are equally exposed to the same source, *i.e.* when by being placed equally *near* the source, all effect of medium is counteracted, and when, as in the theory, only size is taken account of. (3.) Even letting the Medium, *i.e.*, the distance in our atmosphere, remain, the screens, by placing the farther one at only thirty inches from the source, and the nearer screen at one inch, afford an easy practical illustration of the fact that, notwithstand-ing the great diminution resulting from our atmosphere, the light at thirty times a greater distance from the source is not as we are told it is, 900 times less. We *see* it is not ; and we see this with an evidence that no opponent denies. (PART II. *in fine.*) The greater distinctness thus given to these points of the theory gives some advantage to the illustration afforded by the screens. Either illustration, however, will enable the reader to understand the theory and the law assigned to it. He will see that, although Light is known not to equalize itself by diffusion, becoming less on each spot, or thinner, as things do which spread (like water, air, or heat), but on the contrary, depends wholly for its position upon the position of its source, and the concentrated or unspread, undistributed, action of this source, yet it is upon this hypothesis of a level-seeking nature for Light, or its capability of being stretched, expanded, and diluted, *independent of its source*, that our professors are encouraged and expected by their leaders to teach this strange Theory for the diminution of Light by the mere Enlargement of the surface illuminated ; some of them, we see, going so far sometimes ahead of their instruc-tions, as to try to teach it even without the Enlargement, and merely by the *unconscious* introduction of a medium and its effect.

Precise definition of the Law and Theory.—The Law then of this remarkable theory is, as scientific men, and even the more enlightened of our professors, always understand it, although not as any of these ever express it, that THE ILLUMINATION OF AREAS IS INVERSELY AS THE AREAS; just as the thickness of the paint upon two surfaces of different sizes is inversely as these sizes. Why, then, it may well be asked, is the statement always made in the more complicated and obscurer form? Why must we say that the thickness of the paint is inversely as the square of the distance? or that the paint becomes less as the square of the distance becomes greater? The distance always meant in such expressions is, as all educated people know, only distance in a pyramid,—a distance that very few understand who are not geometricians; and since the areas in the pyramid become greater and greater as *this square of their distance* from the apex·or source of size, this expression is used in Geometry to define the size of all areas,—even of such as are not actually in any pyramid, nor ever were in any pyramid. But why transfer the expression to Paint and Light? Instead of saying that these when spread over areas are inversely as the areas, or that they decrease, upon each spot, as the areas increase, why should we say that they " diminish as the square of the *distance* increases"? (*See* Appendix, No. 3.) What distance? One naturally asks, Distance from what? and one naturally supposes distance from the source of the Paint or the Light; whereas no such distance is meant;—no other distance being ever here intended than distance in the pyramid or other geometrical figure,—distance from the angle of the cone or pyramid,—from that which causes the sizes. The expression, therefore, is evidently not intended here to make matters clearer, especially when it is introduced as follows :—" Light diminishes in intensity as we recede *from the source of light*. If the luminous source be a point, the intensity diminishes as the square of the *distance* increases." It is difficult to suppose all these writers and Lecturers so utterly ignorant and unintelligent as to imagine that the distance here intended, is distance from the source of Light, instead of from the Source of Size (for distance from the Light does not make surfaces larger), and yet, if these Lecturers see it is not that, it becomes equally difficult to suppose that they make their statements to this effect, with a constant reference to the distance from window, lamp, or sun, or from some one point in them, for the mere purpose of supporting a theory, and of deceiving their readers as to the distance intended in this so-called law of the Inverse Squares, and in the geometrical phrase " square of the distance." But I leave it to others to find out the motive of the language here employed. It is enough for my purpose, to have pointed out the extraordinary misapprehension connected with it, and to have reminded the reader that, as everybody knows, distance from a luminary does not make one surface greater than another, while, according to this theory, it is the larger surface that " dilutes " or diminishes both Paint and Light. Distance in the pyramid makes these larger areas. This everybody can understand ; but everybody can understand also that distance from a luminary does not do so.

THE ILLUMINATION OF AREAS IS INVERSELY AS THE AREAS. Such then, in simple language and shorn of all disguise, is the substance of the Law and Theory now under consideration, the utter unreasonableness as well as unscientific character of which Law and Theory it is the object of the following pages to expose ; and this exposure, which many will no doubt think here abundantly effected in PART I. of the Treatise, by merely a clear statement of what the Law means, and of what the Theory means, will, for the sake of others, be found effected more in detail in PARTS II. and III.

The great principle, then, of this Law and Theory, (viz., that the illumination of areas is inversely as the areas,) is that surfaces, equally exposed to the source, are illuminated upon each portion of them, in the *inverse* ratio of their magnitudes ;—*i.e.*, not more and more as the surfaces illuminated are greater, but contrariwise, and in the sense most opposed to the geometrical principle, viz., less and less as these surfaces are greater; when the surface is large, the light upon each spot of it is less; and, when the surface is small, the light upon each spot of it is greater; when the enlargement is double, the light is, in this way, reduced to one-half; and when the area is enlarged 900-fold, the light upon each spot of it is reduced to the 900th part of what it was upon each spot of the smaller area; and the advocates of this theory fully recognize, in it, the principle that unless there is this enlargement of the area, there cannot be the diminution that they speak of in the Light, to however great a distance the area may be removed from the source, *i.e.*, from the lamp, the window, or the sun.

Distinction between the geometrical Law and the alleged physical one.—Such is the so-called "Law of the Inverse Squares" freed from all equivocation; and it will be found worth while to keep it well and carefully separated, in our minds, from the geometrical proportion of distance connected with these areas in the pyramid,—a proportion which our theorists append to it, or rather incorporate with it, so erroneously and misleadingly, and as if these proportions of distance in the pyramid had or could have anything in the world to do either with the size of the bodies around us (the walls and pictures for instance), or with the light which these areas receive from the lamp, the window, or the sun. No geometrician, who had even a very slight knowledge of physics, could make this confusion. It belongs entirely to the physicists, and to such geometricians as think themselves also physicists. These tell us, and encourage others to tell us, that their law of "the Inverse Squares" is merely the geometrical law of areas and distances "optically stated;" by which they mean, applied to the illumination of areas; whereas it is nothing of the kind. It is neither this geometrical law applied to Optics, nor is it in any sense this geometrical law at all. *In the first place,* their law of the Inverse Squares is so far from being the geometrical law of areas and distances applied to Optics that it is diametrically opposite and contrary to the geometrical law; and these writers themselves know this, and acknowledge this (although, to all appearance, unconsciously) when they admit that it is the very reverse or inverse of that law,— that it is that law inverted. Surely the inverse of a law is not the law; surely if anything is the contrary of it, it is this. The geometrical law is that A increases when and as B increases; whereas their law is that C decreases when and as both A and B increase. This, it must be allowed, is a curious sort of identity. *In the second place,* the geometrical law of areas and distances shows how any two objects whatever of different dimensions which we see around us in art or nature, can be placed in an imaginary pyramid, so as to determine their distance from its apex, and can thus have their sizes defined by reference to this distance: whereas this size of the objects around us does not determine their distance from the window, the lamp, or the sun; nor can this size be either inferred or defined by their distance from a source of Light. Their size is not in any way influenced by their position in that respect; which is precisely what does happen in the geometrical diagram and in the geometrical law of areas and distances. It would be difficult to imagine anything more preposterous than the notion that this law of the Inverse Squares is an application to Optics of the geometrical law of areas and distances, or that there could be any such application. We have here, on the contrary, this theory that bodies are illuminated in an inverse ratio to their sizes, applied and exclusively applied to the geometrical diagram and the sections of a cone just as if no other application could be found for it, or had ever been heard of for it; whereas the geometrical diagram is but an hypothesis,—what does not exist at all *in rerum naturâ,* and what is intended merely to show how

the sizes of bodies there *determine* their relative distances,
and how these distances, therefore, can define these sizes.
What, we may ask, is there like that in Optics? and what
can be more preposterous than to apply Optics to that?
Thus applied, the theory, already thoroughly absurd in its
principle, works itself up practically, as might be expected,
into the rankest nonsense. How can we speak of different
sizes or areas in Nature as determining their distance from
some apex in Nature whether that apex be luminous or not?
Or how can we speak of different distances from it as determining different sizes? There is nothing of the kind in
nature. Two areas of different sizes are presented to us,—
two books, two tables, &c.,—without any distance known
or thought of either from a lamp or from a corner; and we
are told by these theorists that the larger of these areas
has less light at each point of it than the smaller, and as
much less as this area is larger than the smaller. That
seems absurd enough. But this does not content these
writers. They say that the reason the larger area (viz.,
the one with the greater square of distance) is less illuminated is because it is farther from the window or the lamp;
whereas by the hypothesis both objects are at the same
distance from the lamp, which lamp is, moreover, here regarded by them as the geometrical apex.

What we really have in this Law of the Inverse Squares
is the Inverse Illumination of Areas, or the Spatial Diminution of Light applied merely to the areas of a cone or
pyramid or other geometrical diagram. This diagram
being given, we are told by the physicist that if there were
a luminary at the apex, each of these increasing areas
would be less and less illuminated according as it increases
in size. It is unnecessary, however, to say that it is not
the distance from a luminary but from a corner, or angle,
that enlarges the areas, whose size it is that diminishes the
light.

The foregoing reflections will help the reader to separate,
and widely distinguish, the geometrical principle from the

alleged physical principle which the physicists have so
strenuously endeavoured to incorporate here with one
another; the former principle being the law of areas and
distances, and the latter being that the greater the area is,
the less must be its illumination on each spot, however far
or near the source of the Light may be.

*The Theory imperatively demands Enlargement of the Object
illuminated.*—We now see clearly, from what has been said,
that if, under this physical theory, we wish to establish the
diminution undergone, in this way, by a given amount of
Light, we must show that it has the enlarged area necessary to effect this divisional diminution, *i.e.*, the diminution
at each point of the area. We cannot, under this theory,
suppose that the light upon a given area can be reduced at
all in its whole amount. This, even under the theory, could
only be effected by the reduction of the source. The Theory
does not profess to account for any other reduction but that
of the divisional amount, *i.e.*, of the amount upon each spot
of the area—upon each unity of surface, as they call it.
This is all the Theory pretends to deal with. Nor can we
obtain even this from the theory, by merely placing the
object or area at a greater distance from the source (as so
many writers think we can), while the area remains unen-
larged. It is, by the Theory, indispensable that the Light,
if it is to be thus diminished, shall also, from this greater
distance or by some other means, obtain a larger area upon
which this same quantity of Light may be stretched, spread
out, and thinned. Otherwise there could be, by this theory,
no dilution or diminution whatever. But, as already observed, if it is in any sense a greater distance which gives
this greater area, it is distance from a source of size,—a
corner or an apex,—not from a source of Light,—not distance from the lamp, nor from the window, nor from the
sun,—since it is distance from a corner or angle only, and
not at all distance from a source of Light, which makes an
area larger or smaller.

And here we see the error of those who think that the

amount of Light upon the smaller screen in our illustrations (Appendix, Nos. 2, 3, 15) could possibly be diminished by this theory, if this amount of Light were merely transferred, with its smaller screen or area, to a greater distance from the source of Light.   Obviously in such a conclusion, there is neither Geometry nor Theory, nor anything else.   The area would, in such a case, remain the same, although the distance would be increased, which is contrary to the Geometry ; and the Light at this greater distance would become diminished (not of course in the *whole* amount, but in the *divisional* amount, or amount *upon each spot)* without the requisite enlargement of the area, which is contrary to the theory, as well as, it must be admitted, to common sense also.

The seven Propositions of which the Theory consists.—And now to give a brief summary of the chief principles of this Theory :—

The *four* of these already mentioned are as follows :—

(1.) Surfaces or areas each portion of which is equally exposed to the source are illuminated on each spot in the inverse ratio of their sizes (*i.e.*, inversely as the square of their distance in the geometrical diagram).

(2.) Surfaces or areas, however large and however small, however distant from the source and however near it, have all, in this Theory, the same degree, amount, or intensity of Light upon the whole of each, from the same source, when all are equally exposed to it ; what is here supposed to undergo diminution being only the light that falls upon the parts.

(3.) The whole of the solar light proceeds, completely unimpaired, to the remotest limits of the system, the rays of which it consists being neither diminished in number nor in force in any part of their path ; it being also equally true that all other light is thus wholly undiminished by distance from the source, when, as always in this Theory, exempt, or supposed exempt, from the action of a Medium.

(4.) A main proposition in the theory is that Light, when it has room, leaves its straight line and *stretches* or *spreads* like air or heat (being capable of this expansion to any extent, but only in two dimensions of space), and thus becomes thinned out or "diluted" by being intermixed with darkness.

Three further less important principles of the Theory are as follows :—

(1.) Light can be measured by yards, miles, and inches, square or cubic, as well as by degrees ; a block of light therefore being as possible as a block of quartz.

(2.) *Each* point in the sun's disc, and generally in every luminous surface, is supposed to throw out a *separate* ray to *each* point of all the space exposed to it, and therefore to each point of each surface or area exposed to it ; there being thus formed a cone with its base upon the whole disc or other luminous surface, and its apex upon the point of space or surface illuminated, as well as a cone with its apex upon some point of the disc or other luminous surface, and its base upon some illuminated surface or portion of such.

(3.) Each point of the disc thus acts quite independently of all the other points, having all its energy distinct from them, and sufficient of it to send off a separate ray to every point of the most distant space.

Applications of the Theory and its Law to the Solar System, both (1) *with regard to the size and position of the planets, and* (2) *with regard to the varieties of relative area resulting from varieties of relative distance.*—It remains that we should now point out how this Law of the Inverse Squares applies to the sun and the planets ; and here, as was to be expected, where the areas and the distances with which we have to deal are not in geometrical proportion, while those which are in geometrical proportion are infinite in number, the most grotesque confusion results.

1. We see for instance the theoretical as well as geometrical inaccuracy of saying that, when an area, 20 times greater than another, is also 30 times farther from the source, the light *on each spot* of this great area is (30×30)

900 times less than that *upon each spot* of the smaller area. There is nothing like this either in the Theory or in its Geometry. Neptune's disc, or surface presented to the sun, is not 900 times greater than ours. It is little more than 20 times greater. According then to this Theory (viz., that Light diminishes on each spot in the ratio of the enlarged surface) Neptune's light ought to be about 20 times less than ours at each point of the area, instead of 900 times less than ours at each point, as, contrary to their own theory, these writers represent it on the ground that that planet is 30 times farther than we are from the centre. And in this comparison with us, where is the geometry between Neptune's area and his relative distance from the sun? His distance is 30 times greater than our distance. The geometrical area corresponding to that relative distance is 900 times greater than the area corresponding to our relative distance from the centre, *i.e.*, than our area ought to be. We thus see that the geometrical areas and distances are not the areas and distances of the planets; nor are they almost ever the areas and distances of any of the objects around us. The geometrical areas are only the squares of the relative distances in a pyramid or cone,— chiefly found in geometrical diagrams or other artificial constructions. No two areas that we chance to meet with in Art or Nature, are in geometrical proportion to their distance from some source of light merely because they are in this proportion to their distance from the source of size, viz., from the apex of their pyramid. The relative distance which belongs to Neptune's area and ours in Geometry, is not that his is 30 times greater than ours, but that it is not even five times as great.

We see, then, that if we wish to interpret correctly this Theory of the Inverse Squares, we must be very careful to understand it of enlarged areas only (not of distances from a source of light), and without any attempt to suppress the fact that there is always, by that theory as well as without it, the same amount or degree of light on the whole

of each area, however great or however small the area may
be,—the diminution pretended to by the theory being only
that *upon each spot*, or " unity of surface," as the professors
term it, not a diminution in the amount *on the whole area;*
and we see that, according to this, the received theory of
the Inverse Squares, thus correctly understood, the larger
of any two planets has always less light *upon each spot*,—
less divisional light, as we may call it,—than the smaller
has, although it is not always the larger that is the more
distant from the sun; and all the planets have the same
amount of light on *the whole* of each, without any regard
either to their enormous differences in point of area, or their
enormous differences in point of distance from the sun.
Thus, according to this law and theory, known as that of
the Inverse Squares, the Earth, although so much nearer
to the Sun than Mars is, has much less light upon each
spot than Mars has, and even Mercury much less than the
Planetoids. Jupiter, also, and Saturn, although so much
nearer to the Sun than Neptune, have each of them, by this
theory, much less light than he has, and even Neptune only
20 times instead of 900 times less light than ours ; from
all which we see at once what an important difference it
makes whether we interpret " distance " as from the source
of Light in nature, or from the apex—the source of size—
in geometry.

2. It will further assist the student of this subject, in
his efforts to understand the " Law of the Inverse Squares,"
so justly called by Faraday " a sad jumble of words," if I
here point out to him another result of it not commonly
attended to, and kept, it would seem, intentionally in the
background by the lecturers, as being, probably, too re-
condite for their auditors ; for it is difficult to suppose that
they themselves are as wholly unacquainted with that
application of the Law as they seem to be.

The Law, as we have seen, is to the effect that if the
distance between the Sun and Neptune is divided into 30
equal parts, the light at the end of the thirtieth part is

900 times less, on each square inch of surface presented to the source, than the light at the end of the first part, upon each square inch of the smaller surface corresponding to that shorter distance, and there supposed to receive it; that is to say, as has been already so fully explained, not that the light is less, or that the solar ray has less force at the greater distance, but that it is supposed at that distance to have, geometrically speaking, a larger area to illuminate, although this, we see, does not always happen, and never in the requisite proportion. But the differences of solar illumination which result from this law, even in the space between the Sun and Neptune's orbit, are by no means limited to the calculation that some parts of that space have 900 times more of this illumination than other parts. Far from it. To any minute extent that we choose to sub-divide the distance between any point of space and the sun, to that extent we can, by this law, multiply these differences.

If, instead of 30 equal parts, we divide the space between Neptune and the sun into 100 of these parts, then the light upon Neptune's area is 10,000 times less than it would be upon the much smaller areas in other parts of that distance; or, if we divide his distance into 1,000 equal parts, the light upon that planet can then, the Law teaches, be stated to be a million times less than the light upon proportion-ably smaller areas existing elsewhere in the system; and, if instead of 1,000 we divide this space between Neptune and the centre into a million of equal parts, then, as the square of a million is a billion, the Light spread upon the planet's disc is found by this law, and admitted by these theorists, to be a billion times less upon each spot than it is upon each spot of corresponding areas in other portions of space; and so on, *ad infinitum*, in proportion to the minuteness of the subdivisions which we make in the distance. It is unnecessary to say that the same applica-tion of the Law holds also for a lamp and any divisions that we choose to make of the distance from it in the room. (*See* further on this point in PART III., Section 3.)

FIFTH SECTION.

CONCLUDING REMARKS ON PART I.

I HAVE thus far endeavoured to explain the Law and Theory wherein Light is described as diminishing in the ratio known as the square of the distance (Faraday's "sad jumble of words"), and have done so with what must appear to many an excessive amount of repetition, illustration, and detail, although not perhaps with such an excess of this as may at first sight appear. The justice of the charge, however, must be admitted, it is clear, as far as such readers are themselves concerned. But I have for a long time tried the effect of this preliminary information in short statements upon professional men, with very little success; and as I now address not only the unprofessional men of science who may have had their attention hitherto less drawn to this department of Physical Optics, but also the general reader, who naturally here requires a good deal of preliminary information, I have the hope that, by a fulness of exposition which I at first did not think requisite, even scientific men who write and lecture will at last get to see what I never found a scientific man who does neither unable instantly to grasp. I am tempted, then, to use, on my own humble behalf, the words once employed by a great man in a similar predicament :—" I am afraid I have given cause to think I am needlessly prolix in handling this subject. For to what purpose is it to dilate on that which may be demonstrated with the utmost evidence in a line or two to any one that is capable of the least reflection?" And again :—" If I seem too prolix to those of quick apprehensions, I desire I may be excused, since all men do not equally apprehend things of this nature, and I am willing to be understood by every one."

It will be seen, from what I have said, that the Theory has been hitherto involved in a large amount of uninten-

tional disguise and equivocation, which it was indispensable
to remove; for unless the reader fully understood the
theory which gives the unexpected results now pointed
out, he could not understand what it is that is controverted
in the ensuing pages.

There is, moreover, in all this the received Law and
Theory for the diminution of Light by the mere enlargement of the space illuminated, as well as in the peculiarities
of the other received law for its diminution by Absorption
in a Medium, something so incredibly superficial and unscientific that I have considered it requisite to give, in an
Appendix to this Treatise, a few statements of both these
Laws and of the first-mentioned Theory, in the *ipsissima
verba* of the Professors themselves. Otherwise the assertions involved are so extraordinary, to say nothing else of
them, that the reader would hardly be able to suppose that
these assertions and Laws are here correctly represented.

And the two Laws, now explained, must be carefully
kept quite distinct by the reader. This is a most important
point to be attended to. They have nothing whatever to
do with one another. The simultaneousness of their action,
as held by scientific men, implies no intermixture whatever
either of the action or of the effects. It will be seen, as we
go on, that both laws involve complete misapprehensions
respecting the facts of nature which they profess to deal
with; but of the two, by far the more important, in this
case of the solar system and its solar illumination, is the
Law of the Inverse Squares. It is in fact this Law upon
which the whole question depends which I wish here to
place before the reader, on the Illumination of the Planets;
for, as will be seen in PART IV., the medium between the
Sun and the Planets has almost no effect in diminishing the
light. This Law, therefore, for the diminution of Light by
the enlargement of the Space, or as it is commonly termed,
by "the law of the Inverse Squares," and the alleged facts
of Nature upon which this law has been founded, comprehend together almost all that it will be necessary for me to

treat of in order to explain the great principle that all the planets of our system are equally illuminated by the sun, although revolving round him at such unequal distances.

But let us first attend to the alleged facts of Nature upon which this law professes to be founded, viz., the deviation, spreading, stretching, or diffusion of Light where there is no medium supposed to be present, and the attenuation, dilution, or impoverishment which is supposed to result in Nature to this luminous essence in consequence of the capacity for this deviation from the straight line, this stretching, spreading, or diffusion, which we thus attribute to it independently of all Medium.

END OF PART I

PART II.

PHYSICAL IMPOSSIBILITY OF THE ALLEGED DIMINUTION OF
LIGHT BY THE MERE INCREASE OF THE SPACE
ILLUMINATED, OR, AS IT IS CALLED, BY THE
LAW OF THE "INVERSE SQUARES."

PART II.

FIRST SECTION.

THE ALLEGED FACT.

THE reader now clearly understands, from what has been
said, that, in this celebrated theory of the Inverse Squares,
it is the enlarged space or area alone which diminishes the
Light; Medium and its action being here entirely set
aside,—not however as things which do not exist, but as
things of which no account is taken in this particular
theory;—and that the Distance of the illuminated area
from the source of the Light has no effect whatever,
except in those cases where this distance is required for
producing this enlargement, which, as we have seen, is for
the most part only in the diagrams of Geometry.

This distinction is one here of the utmost importance. It
is only the enlargement or diminution of the space, or surface to be illuminated,—the greater or lesser area alone,—
which has to be attended to, as that which diminishes or
increases the Light. The distance of any area from the
source of the light has nothing whatever to do with this
diminution. (*See* pp. 20, 23, 59, and 61.) The enlarged space

or surface, it is true, is connected in Geometry,—but only in Geometry,—with distance from the source of Size, *i.e.*, with the greater distance at which the space or surface is placed from a given angle in the diagram—not at all with any distance in Art or Nature, nor with any distance from the source of Light. In Art the greater or lesser surface results from the tools or will of the workman, and in Nature from the laws of Nature. In neither of these cases does it result from the laws of Geometry, nor depend on its distance from a source of Light. It is alone then this *enlargement* of the space illuminated, whether it be far from the source of light or near it, and not *at all* its *distance* from this source, which, by this, the received theory, attenuates, dilutes, or diminishes the Light; just as the diminution or contraction of the space or area has, of course, by the theory, the opposite effect.

As far as the geometrical ratio of areas, or angular source of magnitude is concerned, the law, it is well to remember, is that, whatever number of times one surface or area may be greater or less, in Nature, than another, their *geometrical* distance, *i.e.*, that in their pyramid or diagram, is the square root of their difference, and their *geometrical* difference is the square of their distance from the apex, be their distance or difference, in Art or Nature, what it may; well to remember that the *geometrical* fact mentioned has nothing whatever to do with the areas or distances of areas, in Art or Nature,—nothing whatever, for instance, to do with the discs or areas of the planets, and their distances either from one another or from the sun. The geometrical ratio in question is not at all necessary, therefore, in this theory, for the diminution or increase of Light upon any area or surface in nature. It has no more to do with the changes of Light, even in this theory, than it has with the changes of the areas or the surfaces around us. The theory only requires that one of the two areas should be greater than the other.

The Light is described, in this the received theory, as

diminishing at the same rate as the area is enlarged—that is all;—as being 4 times less, upon each spot, when the area is 4 times greater—9 times less when the area is 9 times greater, and so on. This is what is meant by diminishing in inverse ratio to the square of the distance. The whole question then here is : Can this mere enlargement of the space bring about this attenuation of the Light upon it ? Has it always, or ever, this effect ? That is the problem to be here examined.

A given amount of Light from a single source (for we all here only speak of a single, central, or unspread source— a point, as it is often called,) falling equally over a given amount of space or surface,—*i.e.*, falling so that all the space is *equally* exposed to all the source,—becomes, we are told, attenuated or diluted, diminished in strength, intensity, amount, or degree on each square inch, when the amount of the space, thus illuminated, is enlarged ; and this impoverishment of the Light takes place, we are further told, exactly in proportion to the Enlargement of the Space. If, as I have just said, that space or surface is doubled, the Light is thus reduced to one-half of its previous intensity on the square inch. If the space thus illuminated by the single source is enlarged 900-fold, then the degree of Light, available for each square inch or other portion of this space, is supposed to be reduced to $\frac{1}{900}$ of what it was, in precisely the same position, before the enlargement of the space.

This is the alleged fact of nature upon which the so-called Law of the Inverse Squares professes to be founded. We have now to inquire : Is this really a fact of nature ? Is it a thing physically possible ?

All the space surrounding any point,—the lamp, sun, or other luminary,—becomes more and more extensive or enlarged, at every increase of distance from that point ; and this whether the point be luminous or not. This is a manifest fact, and undisputed.

The Enlargement of this space is found to coincide, geo-

metrically, with the square of the distance from the centre,
whether, as I have just said, this centre be luminous or
otherwise. This also is undisputed. The whole space around
the lamp, or other central point, becomes greater and
greater in the ratio of the distance squared; but the objects
and surfaces lying in that space do not become greater
from being at a greater distance; nor does any one given
portion of that space do so. The whole sphere of distance
does, and even the whole hemisphere does. It is obvious,
however, that an area of 20 yards square does not become
enlarged by being placed further from the centre, whether
lamp or Sun, and that even the smaller hemisphere could
not undergo this enlargement, if we could remove it to a
greater distance from the central lamp. (The words here
deceive very few.) And so, of all the bodies occupying that
distance from the centre. They do not become larger
merely by our placing them at a greater distance from the
Sun, lamp, tree, or other centre. The surfaces occupying
that distance,—occupying the space at that distance,—bear
no relation of this kind,—no relation, in fact, whatever,—
to their distance from the lamp or Sun, or other point,
which we may think of as a centre. This is a curious oversight (but not one here of any importance), commonly made
in the geometrical pretensions connected with this subject.
The bodies in question have all sorts of sizes irrespective of
their distances from lamps or other luminaries. The theory
therefore, as it is commonly expressed, cannot even apply
at all to these bodies, but only to the whole circumambient
space (the sphere of space) at each distance around the
luminary, or, at least, to the hemisphere,—that half of this
circumambient space, presented to the disc of the lamp or
of the Sun. The oversight I allude to, however, is not the
disputed allegation which we are considering in the received hypothesis. This disputed allegation is to the effect
that just in proportion as any one sphere, hemisphere, or
other area is enlarged, the Light, which remains in amount
the same as it was at first, and all of which, by this theory,

falls upon all the enlarged area, as it did upon all the
smaller one, becomes thereby impoverished, weakened,
diminished in amount or intensity upon each portion, and,
as it were, thinned out or diluted from being thus spread
out, or expanded over a larger surface than before; and
that, according to their different areas, there is this difference of Light upon each point of all bodies or surfaces
illuminated from the same central point, whatever be their
distance from it.

This is the proposition of scientific men which will be
found controverted in these pages, and easily shown to be
not only utterly false, but manifestly so, if the mere explicit
statement of it has not already had that effect for the
reader. The real question, I repeat, here at issue, is this:
When the source of the Light upon two unequal areas is a
unit or central point (this is an indispensable condition), and
the same unit in both cases, and the whole of both surfaces
equally exposed to it, is the uniform illumination of the
larger surface or space, thence resulting, less strong, less
intense, than that of the smaller one? There is no dispute
as to whether there is the same amount of Light upon the
whole of each area. All recognize that there is. The question is only as to the light at each point, as to the sub-
divisions of the area, and as to what we may call the
the *divisional* amount of the Light. Does, *cæteris paribus*,
the mere increase of the space diminish the amount of
Light upon each spot, leaving it in all cases the same upon
the whole surface? We are here speaking, as I have so
often reminded the reader, only of Light proceeding from a
point, from one concentrated, unspread source. For no one
denies that when we can subdivide and distribute the source
of Light, we can to any extent subdivide and distribute
the Light it gives. The theorists assert that, without any
subdivision of the source, there is always, with an increase
of equally exposed space or surface, a commensurate
decrease of illumination or of light upon each part of the
surface, without any decrease in the total amount of the

Light. We now ask: Upon what grounds do they assert
this? and such is the whole question to which we here
require to devote our attention.

What do they mean by Light thinned out and "diluted"?
What do they mean when they tell us that Light can be
reduced in quantity or intensity at each point, by being
spread out over a larger surface; just as butter becomes
less, at each spot, or thinner, when spread over a larger
piece of bread? How can Light be "thinned out" or
"spread out" at all? Upon what grounds do we suppose
it either elastic or malleable, or inclined, like water, to seek
its own level, or self-expanding? Does it not go *direct*
from the whole source *in straight lines*, and *only so*, to each
point of each surface presented to the source? What is the
meaning of this transverse spreading, so much inculcated as
a scientific fact? How is it effected? Can a ray, without
a medium, be diffused and spread? Can it thus leave the
straight line to deviate into places not in its path? or has
it some expansive power? Such are some of the questions
which these theorists have to answer, and which they have
nowhere answered, nor even tried to answer.

The great difficulty which the reader will here experience
is probably that of bringing himself to believe that the
frivolous statement, now indicated, can be what they mean;
and that of not being deterred by its extreme frivolity from
thinking that scientific men,—men, too, placed high, many
of them, in the confidence of the public,—can possibly be
trying behind their technicalities to teach anything so
obviously incompatible with common sense and nature; and
this merely because they have once committed themselves
to the blunder. They fortunately, however, recognize can-
didly enough, when it is put to them, that this is what they
do. Their hypothesis is simply that just stated (it cannot
be stated too often) to the effect that if we have two unequal areas, the luminous force being the same in both
cases, and both areas equally exposed to the whole of it, the
larger area will have, upon each portion of it, a lower

degree or amount of Light from this same source than the smaller area, and this merely because it is the larger (just as the same bit of butter is thinner upon the larger bit of bread, or as heat and water spread seeking their own level), and that the outer or more distant sphere of space imagined around the sun or other luminary is in this predicament of diminished light merely because it is greater than the inner one ; it being solely on account of its being greater, not at all on account of its being more distant, that it has this smaller amount of illumination,—this lower degree of Light, —upon each square inch of its more extensive surface; although it is quite true that it is because the concentric sphere, thus imagined, is more distant, that it is the greater. This distance, however, is not the reason why discs and the other surfaces around us differ in size.

SECOND SECTION.

THE FACT AS IT IS IN NATURE.

WHEN this, the received theory, is once understood, the fallacy involved in it is of course instantly manifest. The only need in this respect is to comprehend exactly what the theory means—a meaning which in the lecture hall has been hitherto most scrupulously concealed, or most unaccountably unknown.

The Light in a room, with folding doors in each of its four walls, is not diminished when the folding doors are successively opened into other rooms, in which other four rooms there was previously no light. There is in such a case no " diluting " whatever, no thinning out or spreading going on, with regard to the light of the first room. The large additional amount of Light which, in Nature, we here see to be the *true* result from the enlargement of the space,—the Light of the additional rooms,—is supplied from the same single source, without any of this thinning or spreading,—

without withdrawing from the centre room the smallest amount of its original illumination. We see, moreover, that in this case, the four additional compartments of space may be indefinitely subdivided, and the Light thus indefinitely increased.

. Or let us present the curved page of a quarto volume to one side of a lamp, and the curved page of a duodecimo volume in the same type to the other side of the lamp, and, on account of our Medium, but solely on that account, at the same distance from the lamp, (for where there is no Medium as in this theory, the distance it is admitted, produces no other effect but that of enlarging the object); and is it not, I ask, rank nonsense to say that the intensity, quantity, or degree of Light is greater on each square inch or on each word of the one page than on that of the other? How can such a statement as this be in these days designated " Science "? and what is the meaning of the fretful impatience so commonly manifested by the profession when the blunder is pointed out? Why all these efforts made to defeat its exposure? In an age of Science and Progress such a state of things seems to call for some explanation.

We thus see then by two very simple experiments and in our most familiar experience, that the fact of Nature here is not at all as the Theory of the Inverse Squares represents it. The enlargement of the space has not in the smallest degree the effect, in Nature, of diminishing the Light. On the contrary we see from one of these experiments, that, abstracting altogether the action of a medium, as in this theory all writers always do, the enlargement of the space has precisely the opposite effect, viz., the effect of increasing the Light indefinitely without any increase of the source and in an exact proportion to the enlargement of the space; also without the smallest indication in the light itself either of thinning out in the first room, or of spreading, or of being diluted. And no wonder; for this spreading of Light is a physical impossibility. Even in this theory itself it is so. The theory declares Light to consist of nothing but straight lines direct from the source ; each of which lines, however much it admits of being prolonged, is held to be necessarily exempt from all lateral diversion, as well as from stretching, expansion, or enlargement of any kind. But this fact of nature, the fact that Light does not spread, is abundantly shown by these theorists themselves, even in that alone which they all write or lecture respecting the umbra and penumbra of the eclipse. There could be no such thing as a shadow anywhere if each ray had this alleged property of expansion and self-equalization, which appears to be merely imputed to it in order to carry on some other theory.

And the above adjunct of our illustration,—the increase of Light instead of its decrease, as a result of enlarged space,—is, it will be seen from the following Section, not the only consideration of the kind, nor by any means the more important, which lies within the precincts of the received theory.

THIRD SECTION.

ACCORDING TO THE RECEIVED THEORY, LIGHT WOULD IN-
CREASE AS THE SQUARE OF THE DISTANCE INCREASES,
INSTEAD OF DIMINISHING IN THAT RATIO.

I DO not seek to show nor do I hold that (through any effect of enlarged space) Light increases directly as the illuminated area does, or, to use the favourite expression, as the square of the distance does, but it is easy to show that, by the theory and the reasoning employed, it is obviously made to do so.

The theorists offer us, for instance, the alternative, with regard to the larger of two surfaces illuminated by the same source, and both equally exposed to the whole of it (*see* Appendix, Nos. 2, 3, and 15), either that there is a *less* degree of Light on each square inch of it, or a *greater* degree of Light on the whole of it, than occurs in the case of the smaller surface. Now neither alternative here represents the truth. But if, contrary to what they themselves do, we accept the second, if we assert that the degree of Light upon the whole of the greater surface is greater than that upon the smaller, as much greater as the surface itself is greater than the smaller surface, what objection have they to allege against it? Clearly none. Reasoning as they do, they can have none.

They point, indeed, to their screens (*see* Appendix, *ibid.*); and although they do not say as much, they nevertheless leave us to understand that, according to their ideas, Light does not converge at all; that it only diverges; that it does not converge, as we suppose, to each point of both screens from each point of the luminous body, giving the whole Light to each point in each converging cone. They do not, I repeat, say that it does not thus converge; but they seem to think it does not, or, at least, leave us to think it does not. They then tell us that the divergent shadow, or shadow occasioned by the divergent Light, and which represents the measure of Light that is upon the smaller screen, may be regarded as also the measure of the Light which, if unobstructed by that smaller screen, would fall upon the space now occupied by this divergent shadow; *i.e.*, that the larger and the smaller area has always the same amount of Light upon it.

The theory of solar radiation, of which this is a part, will be fully discussed further on (in the next Section, under the Divergence Theory). It is sufficient here merely to observe, (1) that Light converges, as well as diverges, as is commonly explained even in our most elementary text-books, and which no one at all acquainted with the subject now disputes (Appendix, Nos. 1, 2, 16, and 20); (2) that it is the convergent cones only, not the divergent ones, that can be spoken of as illuminating the *points* of any given surface; and (3) that the number of convergent cones which, in this case, fall upon the screen which throws the shadow, is not the number of those which, if not obstructed by that smaller screen, would fall upon the space occupied by the shadow, but, with marvellous exactness, precisely one-fourth of them; from which we see the inaccuracy of what is here " explained" to us (some say " demonstrated ") with so much *naïveté* by the " experiment" of the screens and a medium.   The Light therefore which belongs to the space now occupied by the shadow on the farther screen, is represented, we see, under this mode of reasoning, as being 4 times greater than the Light which falls upon the screen which throws the shadow; in other words, Light increases in the ratio which we speak of as the square of the distance, *i.e.*, in the ratio of the area illuminated.

As the general reader ought to make himself well acquainted with this part of the subject, it may not be inappropriate that I should here give another illustration of the curious fact that, upon the theory of these writers, and from their own reasoning, it necessarily follows (contrary to their intention) that, from the same source, the greater surface has always the greater Light, and as much more Light than the other surface, as it is itself greater than that other, or (in their geometrical phraseology) as it is the square of the greater distance.

Let us take three separate surfaces, each 1 foot square. These, being equal, have, by the theory, an equal degree or amount of Light on each, however near or far from the source they may be, as they belong geometrically to the same distance ; and, by the theory also, whatever we say about them has no reference to the action of a Medium. Let us then unite two of these three equally illuminated areas ; so that we now have but two surfaces or areas, one twice as large as the other.   Then, as this theory which we are examining *adds together the illumination of the parts to find the illumination of the whole surface*, the larger surface, being twice the size of the other, has, by this theory, twice the Light that the other has.   If now, further, to the larger one we add successively, and within the same influence of the same source, 7 more square feet of surface with the amount of illumination belonging to each, and which is, by the theory, the same on each of them, as on each of the other square feet already employed, then we have two separate areas or surfaces—a square yard and a square foot of surface—the square yard, by this theory and by this mode of calculating Light, having 9 times more Light than the square foot. There are here thus two unequal surfaces, with an amount of Light, upon each *ninth* portion of the greater surface, equal to the amount of Light upon the whole of the smaller surface; and therefore, by this mode of calculation, with an amount or degree of Light upon the whole of the greater surface, 9 times greater than that upon the whole of the smaller: which is an amount in exact and direct proportion to the enlargement of the space or area; *i.e.*, to its greater size, and therefore to the square of the geometrical distance belonging to these areas.

Such is their own reasoning, and such the result of it. According to their assertions (*see* Appendix, *ibid.*) the larger space or surface has always, on the whole of it, the same amount of Light as the smaller has, but less Light on each " unity of surface," as they quaintly express it, than there is upon the smaller surface. In the above example, however, we see, on the contrary, that according to the *reasoning* employed in the theory, the two unequal surfaces have neither of these peculiarities. The larger surface, merely from being larger, has, by this theory, the greater Light upon its whole extent; and the smaller surface has precisely the same amount upon the whole of it as the larger one has on the same " unity of surface," viz., on one-ninth of it; the natural inference from which presents these theorists with the dilemma, that either the greater surface has more Light upon it than the smaller has, in proportion to its greater area, or that it has the same degree or amount of Light upon each point of it as it has upon the whole. It is for this latter proposition that I here contend; and it is this which certainly all would hold, except those who advocate the Inverse Squares.

In considering this matter, let the reader be carefully

upon his guard against all language calculated to make him suppose that there is here anything to be taken account of except the comparative size of the surfaces, or, as it is called, the square of their relative distance in the diagram.   The term "distance" is, as has already been explained, very deceptively employed upon this subject. It is the size only, by this theory, or, which is the same thing, distance from the source of size, and not at all, as in a medium, the distance from the source of light, that makes the Light more or less. The distance from the source of size, or geometrical proportion of distance in the diagram, which people here unconsciously speak of so much, makes no difference whatever except difference of size.   This distance is only spoken of, in their theory, as what can fix a geometrical proportion between the greater and the smaller surface, when there is nothing else to fix it; and so furnishes a mathematical term, with its flattering air of certainty, for the expression of the law; this law, however, merely being that the greater and the smaller surface have the same amount of Light on the whole of each (the space occupied by the shadow, for instance, on the larger screen, and the smaller screen itself which throws the shadow), but that each portion of the smaller surface has more of this Light than the same portion of the greater, according to the difference between the sizes of the two surfaces, upon the bread-and-butter principle of the theory, as already explained.

This is the whole theory of the Inverse Squares.   According to it, when one surface is twice the size of another, the smaller surface has the same amount of Light upon it as one-half of the larger surface has, while it also has the same amount upon it as the whole of the larger surface has; which shows at once that the whole Light is on the whole and on each portion at the same time.   But passing over this obvious fact, the advocates of the theory take refuge in a complication, and say that the Light on each square inch or foot is here on the greater surface $\frac{1}{2}$ of what it is on the smaller; when 4 times the size, then $\frac{1}{4}$ of what it is on the smaller; when the greater surface is 20 times times greater than the smaller (as Neptune's disc is than the Earth's), then each square inch or mile of the greater surface has $\frac{1}{20}$ of the Light which the smaller has on the square mile or the square inch, and so on. This is the whole of the physical law here supposed to be true of Light. The mathematical addition has, any one can see, nothing whatever to do with the Light; and is merely to the effect that, in a geometrical diagram at twice the distance from any point, the area or space enclosed between lines diverging from that point becomes 4 times greater; that, at 4 times the distance, it becomes 16 times greater; at 30 times the distance, 900 times greater; this relative area always being the square of the relative distance. What has to be proved, however, by these theorists, respecting the Light, is not this law of areas, which no one disputes and which has nothing to do with the Light upon the areas; nor yet have they to show that, *cæteris paribus*, the amount of Light upon the greater surface or area is precisely the same as that upon the smaller; for this also all are agreed about. What they have to prove is that, this being the case, the whole Light falling upon a surface from a source to which each part of it is equally exposed, is divided and distributed among any number of parts into which we may choose to divide the surface, although the source is not, in any sense, divided or distributed; that although the whole surface is equally exposed to a given Light, each part of the surface is not; that the whole degree or amount of Light upon a surface consists of the Light of the several subdivisions effected in the surface, precisely the same as if there were this division and distribution of the source;—that if one area is double the size of another, and the Light upon the larger equal to that upon the smaller, the Light upon the whole of the larger is not equal to that upon half of it; and further, what has to be made quite clear by these writers and lecturers is, that the received theory of thus distributing the Light of any surface among all the possible subdivisions of that surface does not represent Light as increasing in the ratio of the areas,—in the ratio which they describe as the square of the distance,—instead of as decreasing in that ratio.

FOURTH SECTION.

THE TWO MAIN SOURCES OF THE ERROR RESPECTING THE FACT OF NATURE IN THIS CASE; VIZ., THE MULTIPLICATION THEORY OF FORCE, AND THE DIVERGENCE THEORY.

So far it has been seen not only that, in Nature, there is none of this supposed expansion, stretching, or spreading out and enfeeblement of Light from enlarged space which is mentioned by the professors, but even that the result of their reasoning is that Light increases, through this enlargement of the space, exactly in the ratio in which they think it diminishes. I again repeat for those who need it, that I am not here engaged in discussing the question as to whether Light does so increase. I only point out that it does so according to the alleged facts and logic of these theorists; and I do so in the hope that this discovery may help them to lay aside this logic and these " facts." I have now to speak of the two erroneous notions, formerly prevalent, which became the principal sources of the strange Law and Theory so long talked of and believed in as that of the Inverse Squares. When an error is so great as to be considered unaccountable, it is, by most people, or at least by many, considered to be either no error or half true. Those, therefore, who point out extraordinary blunders are generally called upon to explain the origin of these blunders, and are told that unless this is done, their disproof of them can have no thorough effect in convincing the less profound portion of those who will undertake to judge. In accordance, then, with this principle, I proceed now to point out the two main sources of the extraordinary blunder before us. For there are two of these sources; one the more plausible and popular, which might even alone have suggested the theory of the Inverse Squares (but without the second could not have had much effect), and which, although itself an obvious blunder, is still as current a conviction among the learned as it ever was. It is, therefore, necessary to analyse it minutely. This we may here call the "Theory of Divisional or Multiplied Force;" which is to the effect that the amount of any Quality existing, or Force in action, upon any given portion of matter is equal to the amount of this Quality or Force which we find in any one of such sub-divisions of this matter as we may choose to divide it into, multiplied by the number of parts into which the mass or surface is thus divided; according to which the degree of light upon a picture 5 feet square, *i.e.*, upon 25 square feet, is 25 times greater than that on one square foot of it, being thus distributed over the whole picture. The other source of the Inverse Squares is the Divergence Theory or Spoke Theory; which teaches that two rays, diverging from a luminous point, gradually augment the space between them more and more, as they proceed farther from the luminous point; and can, therefore, do less and less for illuminating that increasing space and the objects that lie within it. This Divergence Theory many will think the true origin of the blunder, and in some respects it can be said to be so, because it was the first of the two asserted with much prominence; but it could not originally have effected anything, nor could now continue to have the weight it has, without the co-opera-tion of the other source. I therefore explain both these notions; and begin with the Theory of Multiplied Force, which, as I think, prepared the way for the other; which, moreover, is still highly popular among professional men, and is in fact all now that gives an air of plausibility to what they teach upon this subject of the Inverse Squares. This is that popular conviction upon which the obscurer and more scientific Theory was able to work,—that, in short, which constituted its basis. Without this Multiplication

Theory, speculative minds would never have been decoyed originally into the confusions of the Divergence Hypothesis.

1.—*The Multiplication Theory for the Forces of Nature (one main source of the Error now being exposed).*

The first misapprehension here found by the Theorists ready at hand to work with,—in which they were, no doubt, themselves fully immersed,—the error already in possession of men's minds in connection with this subject, was the belief that when we divide a surface into any number of equal parts, we thus divide also, and lower to that extent, the degree of Light, Heat, Colour, or other Quality, *i.e.*, the effect of any Force which, it is admitted, there is in action upon the whole surface taken together; and that we must, therefore, multiply this reduced degree of Light, &c., existing upon each part, by the number of these parts, in order to find again the degree of Light or other Quality upon the whole surface.

It has been fully acknowledged from the first by most of these speculators, that every surface whose parts are all equally exposed to the whole Light, has the whole Light falling upon it, whether this surface be great or small; and about that there is now little or no dispute. Some indeed have written as if they imagined that a surface could be *exposed* in all its parts to the whole light of the solar disc without having the whole of this light *falling upon* it; but this seems to be in consequence of a misapprehension as to what is meant. Such writers seem to imagine that by the expression " the whole light of the solar disc," we mean, not the amount of it experienced anywhere, or anywhere existing, but, on the contrary, an amount of it which exists nowhere in Nature, and therefore can be nowhere experienced; the amount, namely, which would result from the application to the sun, and pretty near him, of a lens as large as, or rather larger than the solar disc itself; by which the whole light would be taken from all the rest of nature opposite the disc, and be all concentrated upon the point of a needle. In this sense, even these curious scientific speculators themselves admit that nothing whatever has the whole light of the disc falling upon it,—that nothing ever is, or ever can be exposed to it. All this then is simply a misapprehension on the part of these Writers and Lecturers. Nobody means this when we speak of anything as exposed to the whole light of the disc, or the whole of this light as falling upon a surface. We only mean this whole light *as it exists in Nature,*—as the natural philosopher experiences it; and they themselves mean nothing else, except for the purpose of evading facts that are adverse to their theories. They themselves admit that the whole light of the disc is upon the whole hemisphere of space, however distant, which is opposite to the disc, which also is something very different from the point of a needle; and the light upon it something very different from this concentrated light of the disc which they call the whole of it. In no part of that hemisphere is there the supernatural degree of it above indicated, and which they always speak of, when it suits their purpose, under the same name as the natural degree of it with which we are all familiar. I have said all this that there should be no equivocation possible; and that it should be clearly understood that the *whole* light of the disc, *as it is in nature,* falls upon every surface which is exposed to the disc, and is by all reasonable people considered to do so; and nowadays, as far as I know, even by all professors. But it will be seen that it is of no importance in this discussion, whether we speak of the *whole* light of the disc (in any sense), or only of some part of it, as falling upon any given surface. The real difficulty which our theorists profess to experience and to deal with in their theory of Diminution by Enlarged Space, is quite independent of the question as to what is, *in nature* or out of nature, the whole light of the disc. Their difficulty is this: If, as so many justly think, the whole light of the sun or other luminary

upon the whole surface, let us divide the surface into

any number of parts, and in doing so, we also, say they, *necessarily divide* the light into this same number of parts, although, it is true, the source is *undivided*. It follows then, they further tell us, that each *part* of the surface has its own *separate* and exclusive part of the light, instead of having the whole light upon it; that the whole degree of light consists of all its parts, just upon the same principle as the whole surface itself, along which it exists, consists of all its parts; and that each of these portions of the light therefore must be less than the whole of it. Or even if the *whole* light does not fall on each surface exposed, yet since the larger and the smaller surface have each the same degree of light from the one source upon them (as all these theorists without exception allow), how can each square inch, they ask, of the one surface have the same portion of this light as each square inch of the other surface, there being more of these square inches to supply with light in the one surface than in the other?

The answer to this difficulty of these scientific men, in both these cases, is the same. This answer is that it would be physically impossible to have any division or separation of the whole light without a division and separation of the source; and we have no such division or separation of the source connected with this investigation. The source of Light here in question is an indivisible, central, unextended unit or point;—such that no one part of it acts without the rest,—and the whole area in question is, in every part of it, equally exposed to the whole source; for which reason the source is called central.

It is seen, from what has been now stated, that there are here three judgments under discussion respecting the comparative illumination of different areas; or, to say the same thing in other words, three judgments respecting the illumination of two surfaces, the one of which is four times, or nine times, or sixteen times, or 900 times, or any number of times, greater than the other; which difference of area is, in a geometrical diagram, described as being the square of

the distance ;—relations of size and distance which, as we
have seen in Part I., do not exist at all among the bodies
of the solar system, and but rarely anywhere in nature ;
yet in this geometrical phraseology it is that these theorists
prefer to describe the area they require to speak of, thus
vaguely bringing in, with the term " distance," the familiar
idea of what happens in a medium, and, with the expression
" square of the distance," giving to the statement an air of
mathematical certainty, which they, not unnaturally, like
their theory to have ; for neither of which attractive inti-
mations, however, is there here the slightest pretext, as has
been already so fully explained, in the portion of this
Treatise just alluded to.

1. One of the three judgments in question is that the
same degree of Light is on the *whole* surface, uniformly
exposed to the source, and also on *each part* of this surface
at the same time and in the same sense, whatever the size
of the surface may be. This is what we consider the conviction of common sense, the plain fact of nature and of
science, and the ordinary judgment not only of mankind in
general, but also of the scientific who are unprofessional.

2. Another judgment is that the degree of Light which
is on the *whole* of each surface is not on each *part* of it ;
there being on a part of the surface only a part of the
degree which there is upon the whole. And if we compare
two unequal surfaces, one of which is four times greater
than the other, then there is, on each part of the greater surface, only one-fourth of the degree, amount, or intensity which
there is upon an equal part of the smaller surface. This
is the notion taught in their books and lectures by all professional men under the name of " the Inverse Squares,"—which
means the geometrical law (of areas), inverted or reversed.

3. Another judgment is that the whole light is upon each
portion of both these surfaces, but 4 times more of it upon
the whole of the greater surface than upon the whole of the
smaller one, on account of its being 4 times greater. This
notion is clearly as contrary to common sense as that in the

preceding judgment is, and is even disclaimed by professional men; but, as shown above (Section 3), it follows necessarily from their own principle; which is that the degree of light distributed over the whole surface is the sum of that upon all the parts.

Now, a capital point to be attended to, in all this, is that, according to the professional doctrine as well as to that of all scientific men, the greater and the lesser surface, whatever be the proportion which these bear to one another, have, on each of them, precisely the same amount or degree of Light. The question at issue relates only to equal parts of both surfaces, or to what are called their "unities of surface."

And, from this, we may also see it to be universally recognized, although not universally expressed, that the whole of the natural light from any luminary,—from the sun's disc for instance,—to which any surface, great or small, is exposed, falls upon that surface; for, otherwise, Light from one and the same source could not be equal, in this way, upon surfaces of all dimensions; and that there is nothing incongruous, but, on the contrary, scientifically exact in this universal judgment, respecting the totality as well as the identity of the amount or degree upon the greater and the lesser area, is proved by the fact that this totality as well as identity of amount is true also and necessarily true of the greater and the lesser of the concentric spheres already mentioned. Even if one of these spheres is a million of times greater than the other, each of them receives upon it the whole of the central light and therefore the same degree of it. This principle is also illustrated (Appendix, No. 15) by placing a taper successively in the centre of two boxes of different sizes, whereby the whole interior surface of each box receives upon it the whole light of the taper, (allowance being made for the effect of medium,) however great the difference of size may be between these boxes. On this point of TOTALITY, although, in this discussion, entirely unimportant, it is useful to remember the fact of nature, and that, respecting it, most professional men nowadays agree with all of us.

What is here important, and what all, without exception, seem now to be agreed about, is that the smaller surface and the greater,—the space occupied by the shadow for instance (*see* Appendix, 2, 3, and 15), and the smaller space occupied by the screen which throws the shadow,—receive, upon each, the same amount of Light from the same source, when equally exposed to it; and that there is the same degree or amount of the sun's light upon Neptune as there is upon Mercury. Up to this point all are agreed. It is here that the professional theory begins. The professors hold that notwithstanding this equality of light upon the two most unequal areas, equally exposed, and, in fact, in consequence of it, the light becomes diluted, attenuated, enfeebled, and diminished, by the greater number of parts over which it has to be diffused and spread out in the case of the larger area.

They consider that the sun's rays undergo no diminution whatever, either in force or number, to the remotest limits of the system. They consider that one of these rays passes from each point of the sun's disc to each point of space, and therefore to each point of each surface occupying space ;— that there is, therefore, a cone of the whole light with its base upon the disc extending to each point of each object, however distant and however near. They consider also, and for the same reason, that the whole of the solar light falls upon every conceivable sphere of space which has the sun as its centre, however distant it may be from that centre; which concentric spheres of space, as has been explained in a former page, necessarily increase greatly in size, as their distance from the sun increases,—much more rapidly than the distance itself does,—and in the ratio, as we have seen, of this distance squared.

There is, then, no difficulty in comprehending that, even according to the theory of the Inverse Squares, the whole undiminished light of the sun covers equally the larger and the lesser sphere; the whole undiminished light of his disc therefore, the larger and the lesser hemisphere, as well as the larger and the lesser space or surface or portion of the sphere, presented to it. All this our theorists fully recognize. They then ask, as I have already observed: But since there is thus only the same degree or quantity of Light for the larger area as for the smaller, on which point we are all agreed, how could there possibly be the same degree or amount of it on each square foot of the larger area as there is on each square foot of the smaller? In other words: If there is upon the smaller surface all the light which there is upon the larger surface, how is it possible that the light should not be thicker and stronger and more abundant upon each square foot or square inch of the smaller surface than upon each square foot or square inch of the larger surface?

Our answer to this is (as already given) that the thing is perfectly easy and perfectly natural. Light is not measured by feet or inches but by degrees; nor does it thicken and strengthen, when the source does not; nor does it thin out or spread until the source does. In physical Optics, when the central, undistributed source remains the same, the effect remains the same, how great soever may be the enlargement or contraction which takes place, or can be supposed to take place, in that which, from such a source, receives the light. One portion of the whole light does not go to one spot, and another portion to another spot; but the whole light to each spot. What seems here, to some extent, to bewilder these writers and lecturers is that they forget the elements of their own theory. They forget that, according to this theory, the same degree of light from the disc falls upon each space or sphere or surface exposed to it, whether that surface be near to, or far from, the sun,— whether, also, it be great or small; inasmuch as there is a ray from each luminous point of the disc to each point of space however distant, and this ray undiminished from the first to the last. Now, this being the case where, as in this theory, no medium is taken account of, and being admitted by all to be so in such a case, how is it physically possible for the mere *size* of the space, exposed in all its parts to the same source, to make the degree of light less or more upon any part of it? If one sphere or area be 20 times greater than the other, the greater will contain 20 parts, each part equal to the whole of the smaller area. If, then, the whole light is equally upon both areas, as all agree it is, how can the twentieth part be better lighted (have more light) in the one case than in the other? when it is by itself, than when it is combined with 19 others? If, as in the instance previously given, there is precisely the same amount of illumination upon an area a foot square, and upon an area a yard square (which latter is equal to 9 square feet) how can we say that the square foot of the larger area has less of the illumination, grouped, though it is, with eight other square feet of space, than when, by itself, it constitutes a separate area which is a foot square? Since the same amount of light falls upon each part of space,—upon each area and each sphere,—how can we arrive at such a conclusion as this? There is no one, I think, who will not easily see the unreasonableness as well as the groundlessness of the hypothesis in question.

If, as already remarked, the source illuminating both areas equally, were not a unit,—a physical point,—a central, unspread source,—the case would be different. If a group of twenty candles, for instance, instead of giving their light thus collectively, as the Sun gives his, are separated and distributed so that each candle lights a separate area or portion of the area, the degree of Light is diminished by the division and distribution of the source, the degree being necessarily much lower from each candle separately than from the group. But this effect is not caused by the greater surface with a single central source of light; which is what we are speaking of; and this subdivision of the source does not occur in the case of the Sun.

But, beside the unreasonableness now pointed out, this

Multiplication Theory is, as already shown (Section 3), utterly opposed to the Law of the Inverse Squares, which it is supposed to justify. Since all admit that a square foot, presented to the sun, has the same amount of illumination, whether it is alone or is combined with eight others, then it would follow necessarily, from the supposed subdivision of the light, that the yard square would have 9 times more light than the area of one square foot has; *i.e.*, that the light from the source would be increased ninefold in the case of the larger area, instead of being the same for the two areas;—a conclusion which, as explained above, entirely subverts the theory of these writers, and represents Light as increasing instead of decreasing, exactly in proportion to the enlargement of the area, *i.e.*, to the square of the distance;—the common, unsophisticated interpretation of nature here being that, in all such cases, there is the same intensity or amount present,—the same luminous force in operation,—upon the whole of each surface as upon each part of each,—the same degree of light upon the square mile and its square foot, when both are, in all their parts, equally presented to an unspread undistributed unit source.

This Multiplication theory for the degree of light from a single source, so evidently the Encouragement, if not the Origin of the whole error, is almost too manifestly foolish to need that I should have said much about it; and yet it has served to mislead some of the most distinguished Physicists, not in England only, but in the other countries of the world; which must be my excuse for what cannot but to many seem a very needless prolixity. The intensity of the report which, when a cannon is fired, extends equally to the whole of two areas of different extent (say, the fortress walls on each side of the gun), is not only precisely as great for the greater area as for the smaller; which we all admit it is; but also for each square yard of each area, as it is for the whole area together, which is here what our theorists deny (Appendix, No. 15). We cannot proceed to add the noise audible on one square yard of the walls to that audible on

another square yard, and then say, after summing up some
80 such square yards, that the whole report was, at least, 80
times louder than on any one square yard of the walls.
Can this be called science? The intensity also or depth of
a Colour which exists upon two areas of different extent, is
not only precisely the same,—as intense and deep,—for
the greater area as for the smaller, but also for each square
inch of each area as for the whole of each. In neither of
these cases does the intensity become divided, distributed,
and diminished, merely from our having recourse to sub-
divisions of the areas; and so also the illumination does
not. The entire intensity or force is upon the entire area,
and upon each minutest portion of it, at the same time.
The same is true of Heat and of Attraction. There is here
no difference except for a theorist. If the theorist chooses
or requires it, the vibrations of the Colour or of the Sound
(things which admit of Long Measure) may be subdivided
and distributed over each square inch of the areas over
which one intensity of colour is seen, and one intensity of
sound is heard; but even if this be true of the vibrations,
it is clearly not true of the Colour itself, nor of the noise
itself. We cannot say, of either, that a portion of its
whole amount goes to one square foot, and another portion
of the whole to another square foot; and so on; and then
that the sum of these portions is the whole depth of the
Colour, and the whole loudness of the Sound. To say that
the amount of the Colour or of the Sound is not the same for
the whole space as for the part, is the preposterous thing here
asserted of the Light; but is what any one, however little
versed in scientific matters, can easily comprehend and easily
see to be, on the part of enlightened men, a mere oversight,
however much the less enlightened may be convinced of it.

Those not committed to some hypothesis will at once
recognize that we cannot accept this Multiplication-Theory
for the Forces of Nature;—that we cannot, with any truth,
pretend to estimate the amount of a Force from a given
centre by subdivisions of the space, every part of which is

equally exposed to its action. We cannot divide the space,
thus equally illuminated by a lamp, into fifty equal parts,
and say that the whole luminous effect, or amount, or
intensity of the lamp,—the degree of light it gives,—is
fifty times greater than it is at any one of these fifty
points; for it is not. Everybody knows it is not. The
same degree or intensity is on the whole, and at each point
of the whole, at the same time. This is really here the
important point to be attended to, and all the more so,
because it is one which the Writers and Lecturers en-
deavour to keep out of sight. The error seems to consist,
to some extent, in a confusion made between some theories
and the Natural fact, by the aid of a confusion between Long
Measure and Degrees;—between the Long Measure, on the
one hand, so preposterously applied to a force or agency of
Nature, in order to prop up an hypothesis, and Degrees, on
the other hand, or things measured by Degrees, as Heat,
Attraction, or any Force naturally is; the Long Measure
referring to a supposed extended Element as cause of
Attraction or Light, and being therefore, like this extended
Element, a mere hypothesis; whereas the measurement by
Degrees refers to the Intensity, Attraction, Heat, or Light
itself, and is therefore, here as everywhere, entirely exempt
from hypothesis. Be that as it may, the whole luminous
Force or degree of Light so called, whatever we may think
of its cause, is not more truly upon the whole surface or
space in question than it is upon each square inch of it;
which obvious fact these theorists are themselves obliged
to acknowledge, when they say and truly say, that,
whether the surface be great or small,—a square inch or a
square mile,—the whole degree of Light to which it is
uniformly exposed, is upon the whole of each extent. We
could as reasonably multiply the degree of solar heat,
experienced at any given hour, by the million hours previously elapsed, averaging the same degree of heat, and
thus consider the solar heat during all that time as a million
of times greater than what we experience at the hour

mentioned, as say that the Light upon a square foot is 144 times more intense, stronger, greater in amount or in degree, than that upon any square inch of the square foot. Or, if in a room 20 feet square, the thermometer marks 70 degrees on each square foot of the floor (to take no account of the body of the room), is it common sense, on our part, to say that the heat of the floor alone amounts to 400 times more than 70 degrees of Fahrenheit? Or, if there were 100 people standing in the heat of a summer sun, is there common sense in saying that the solar heat is, in this case and on that spot, 100 times greater than any one of these people experiences it? Would not that be simple nonsense? Yet this is, with regard to Light, what these Writers and Lecturers inculcate. Or would it be reasonable to say that because the blue of a given yard of ribbon is very deep, this colour is 100 times deeper, in 100 yards of this same ribbon, than in one yard of it? or that, when we divide the surface of a mahogany table into 100 equal parts, the hardness of the whole surface is 100 times greater than that of any one part? or that, because a man who has once shouted as loud as he can, does this 100 times, he therefore shouts 100 times louder than he can? It may be well that all these Writers and Lecturers should reflect a little upon the utter nonsense they thus inculcate.

In all these cases the error made seems to be, as already observed, a confusion between the cause and its effects; a confusion between Light and the supposed causes of Light, between Heat and the supposed causes of Heat, between Colour and the materials or vibrations with which it is produced, &c. This confusion has led these writers to suppose that, when a larger and a smaller area are lighted, heated, coloured, rendered noisy, &c., by a common or central cause, *i.e.* by one to which each portion of both spaces is equally exposed, there must be a greater degree of light, heat, colour, noise, &c., upon the whole area than upon a part of it. Whereas this is never the case. If there is this single or unextended cause, this central source or point, as it is often called, *i.e.*,

one to which the whole of each area is equally exposed,
then the degree or amount of effect produced has no reference whatever to the amount of space over which it is produced. There is precisely one and the same amount of
effect produced at every point in the case of both the spaces;
and no reasonable person ever thinks of calling the amount,
at each point or square inch, a separate effect, and of adding
these effects all together, in order to find the *real* amount of
noise or colour upon the whole of either space, nor of saying
that the effect is greater upon one of these two spaces than
upon the other.

If, on the contrary, the source or cause is not common,
but divided or extended, *i.e.*, manifold,—if that which produces the uniform effect on one portion of the space is not
that which produces it upon another portion,—then the cause
in action upon the larger space must be as much greater than
that in action upon the smaller, as the larger space itself is
greater than the smaller, and the cause in action upon each
separate portion of each space must be less than that in
action upon the whole of each. To produce one uniform
effect upon 20 square feet of surface in such a case of
divided or extended source there must be 20 times as much
of this source or cause as would be necessary to produce
the same effect upon one square foot; and thence arises the
confused judgment that even the effect, instead of being
the same on both spaces, is 20 times greater in the one than in
the other. The confusion here, however, is evident. Even
then the effect *here in question* is not made less or greater
by this less or greater amount of source or cause. Precisely the same effect extends over the larger as over the
smaller surface.

But when there is not the excuse of a divided and uncen-
tral source, for this strange interblending of disconnected
ideas on the part of men accustomed to the precision of
scientific research,—when there is but the one undivided and
central source to be taken account of, and this, giving precisely the same effect for the small surface and for the large

one, we easily see the unreasonableness of saying that it produces a higher degree, amount, or intensity of its effect (whether this effect be noise, heat, light, or colour) for the larger surface than for the smaller, or (if possible still more preposterous) for the smaller than for the larger.

It is true, for instance, that the cause of heat for the 400 square feet of surface, in the instance at p. 82, must be 400 times greater than for one square foot. Of this there is no question; for each of these square feet cannot be equally exposed to the action of the source which warms one of them. If this were possible, then the source of heat for one would be sufficient for all. But as things are in this supposed case, although the source must be greater for 400 square feet than for one square foot, the heat itself, —the degree or amount of heat,—must not be greater for the whole floor than for one square foot of it; and so also of Light. There may be the same intensity, amount, or degree of light on a space of 20 square feet, as on a space one foot square. If these two unequal surfaces can be so placed (say, round a lamp) as to have each square foot of the larger equally exposed to the source of light as the one square foot of the smaller surface, then, no more source is required for the greater than for the lesser surface in order that precisely the same degree of light should be produced. If these two unequal surfaces cannot be so placed as here described, then it is necessary to give each square foot its cwn separate source, this source therefore being, in the case of one surface, 20 times as much as in that of the other. In like manner, although the blue in a square yard of surface be exactly the same as that in 100 square yards, —the same intensity, the same shade, in short, the same colour,—yet the source of this colour and its depth or quantity in the one case (be this source materials or vibrations) must be 100 times more in size,—more extended,— than in the other case, unless we employ a central or unextended source, i.e., one whose efficacy does not depend upon its extent or number; as, for instance, a central blue

light. So also in Sound. In all these cases it is clear that
the unextended or central cause can produce its effect over
any extent of surface equally exposed to it, and this, without either the effect or the cause being subjected to any
supernatural multiplication in the process; and that where
the concentrated or unit-cause cannot do this, it is because
there is none,—it is because the extent of surface, not
admitting of being equally exposed to the unit-action,
requires a separate cause or source for each portion of it,
and therefore as many times more of this cause for the
larger surface as this larger surface is greater than the
smaller. But even then, as I have said, the effect here in
question,—whether it be Noise or Colour, Heat, Attraction,
or Light,—does not become multiplied or augmented by this
fact. It is not more intense, not greater, not stronger, for
either surface than for the other, even after all these additions to the cause. We see, then, here as everywhere, the
importance of carefully guarding against the least confusion between the cause and the effect in nature,—a
confusion of ideas often so much required for theoretical
purposes;—and it is hoped that, from what has been said,
the reader will have no difficulty in clearly recognizing the
two main facts connected with the confusion in the present
case:—(1) that we cannot divide and distribute the effect of
an unextended or central Force, because we cannot divide
and distribute the central Force itself; and (2) if we destroy
this centrality (or absence of Extent and Number) by subdividing and distributing the Force, we do not thereby
augment the effect. We do quite the reverse. We do not
even preserve it. We thereby reduce at all points, and to
that extent, the intensity, amount, or strength of the effect
produced; which is what our opponents suppose done, even
while they retain the central Force.

There are few, I feel confident, even among the most
prejudiced of our professors, who, however silent they may
choose to be, will not be able here to recognize the equivocation and confusion into which they have hitherto been

decoyed upon this subject of Cause and Effect in Nature, and the further confusion to which that leads respecting the substitution of Long Measure for Degrees, and the mere Bulk for the Action of "the vibrating mass unseen" which is supposed to be one cause of Light; as well as the third confusion, resulting from this latter one, and which has led so many to imagine that the degree or amount of Light upon a surface is to be regarded as something made up of all the degrees or amounts of it supposed to exist separately upon all the sections into which the surface is divided or divisible.

2.—*The Divergence Theory (the other main source of the Error now being exposed).*

The second of these two theories which have chiefly contributed to the Misconception by which the enlargement of the space illuminated is supposed to diminish the Light falling upon it, is commonly known as the Radiation of Light, but is more properly that peculiar form of it which may be called the "Theory of Divergence" or the "Spoke Theory" (Appendix, Nos. 1 and 3), also the "Radiation of Darkness," or the "Theory of Unilluminated Cones," a theory which supposes more than one-half of all the space around the sun, however near him, to be in utter darkness, and which makes the sun invisible, as well in parts of the system near him as in parts at a distance from him. This theory, it will presently be seen, is to the effect that Light does not leave the sun in a sheet or flood, as the Force of Gravitation leaves the Earth or the Sun, nor yet in rays laterally continuous, but, on the contrary, in disconnected rays,—in fine lines or threads, gradually separating more and more from one another, and thus forming a cone of Darkness, between every two lines, along their whole length,—a circumstance which would have (it is rightly supposed by its authors) the effect of preventing large portions of space from ever receiving any of the sun's rays at all, however near him these portions may be, or however

much exposed to his whole disc. It will also be seen that
for this hypothesis there is not, in Nature nor in reason, the
smallest foundation. It is merely one of those now commonly employed as an explanation for the supposed diminution of Light without the action of any Medium, but so
utterly baseless in itself as to make one suspect that it has
been invented expressly for that purpose, were it not known
that it has been itself to some considerable extent the origin
of that notion. It will be well, therefore, to look a little
further into this hypothesis also; of which, however, one
sees with pleasure that some of the theorists themselves
seem already beginning to be ashamed.

There are two ways in which (as well under the unfor-
tunate Emission-hypothesis as under the more prosperous
hypothesis of Undulations) Light, when exempt from
Medium, can be supposed to *exist* in empty space or to *pass
through* it; viz., either in a connected flood as gravitation,
water, and air exist and pass, or in disconnected threads,—
infinitesimally fine lines,—called rays. In the first of these
two forms of propagation scientific men hold that all diminution of Light, without a Medium, would be impossible.
They are therefore, we find, compelled to resort to the Ray
Theory to account for the reductions theoretically supposed
to take place without the action of a Medium. The reader
should distinctly understand this emergency. If Light
existed as a sheet or flood or aeriform mass, and proceeded
thus from the solar disc as the Force of Gravitation exists
in space and proceeds from its centres (there being no Ray
Theory for Gravitation) professional men admit that there
could in such a case be no diminution of the Light independently of a Medium; and assuredly, it is not easy to see how
there could then be any, unless upon the same principle as
Gravity diminishes; viz., because a certain amount of force
or action belongs to a certain distance. The solar light would
in such a case, they all admit, pass on in all directions, undiminished, to the limits of space; although they recognize,
it is true, that, notwithstanding the absence of all ray

theory, this uniformity throughout its path does not occur in the case of Gravitation—nay, assert that the same law of the Inverse Squares applies to Gravitation also without any of this Radiation theory. It is not, therefore, in this continuous sheet or flood that the luminous effect is supposed to exist in space, apart from the Medium, proceeding in one piece, as it were, from the whole breadth of the luminous body. On the contrary, it is supposed to *leave* it, as I say, *in lines*,—in fine threads of light called rays,—with equally narrow spaces of darkness between them, and to *exist* in this form everywhere throughout the system. It is not found that Light could in any other way be *scientifically* diminished. It is not, however, in every form of this ray theory that the required diminution can be obtained.

There are two different ways in which the hypothesis of lines or rays of light can be supposed to exist in space. The more obvious and natural of the two (if either of them can be called either obvious or natural) is that the light diffused around us, and therefore *all* light, consists of these fine rays,—innumerable luminous lines of light,—infinitesimally attenuated, which upon leaving *each* point of the luminous body, diverge from one another in that point at every conceivable angle *so as to be able* to reach *every* conceivable point of space : this Universality of the rays being the very ground and object of the alleged Divergence. According to this form of the hypothesis, one of these rays proceeds from each of these atomic points in the solar disc or other luminary, to each atomic point of exposed space, and therefore to each atomic point of exposed surface, throughout the whole system if not beyond it ;—more than one could have well expected as the work of one single point. In this way *each* minutest point of every surface presented to the sun has upon it *one* ray,—the extreme point of a fine line of light,—derived from *each* minutest point of the disc (however great or however small the distance may be between this point of the disc and the object), and therefore has upon it (*i.e.* upon each minutest point of every surface) the

whole degree of light that the disc can give; for there is
no reason assigned, nor assignable, why one of these
exposed points of space should have rays on it, and another
not; nor why each point so exposed to the whole disc
should not have its full complement of rays upon it, *i.e.*, a
ray upon it from each point of the whole disc. The result
is that no point, or other "unity of surface," has either
more or fewer of these single rays upon it than all the rest
have,—the full degree of solar illumination being thus
brought to each point; nor more when near the sun than
when distant from him; for apart from all Medium, as in
this theory, each point of each surface, and each point of
space, are as completely presented to each point of the
whole disc when they are distant from it as when they are
near it. Mere distance alone, it is clear, does nothing, in
such a theory, to diminish the number of the solar rays
falling upon each minutest point of space or surface that is
exposed to the rays of the disc; and nothing to diminish the
number of these minutest points which, in space or surface,
are exposed to these rays; nothing therefore to diminish
the solar illumination existing at each point in the most
distant parts of the solar system, or existing upon any surface within that system, provided that each point of each
surface be completely exposed to every one of these supposed threads of light, *i.e.*, to all the disc from which they
proceed.

It is manifest that in such a theory we have *converging
cones* of Light as well as *diverging ones*. Let us attend a
little to this fact. We here find, as has been just explained,
cones of Light, indefinite in number, with the apex of each
in some atomic point of the solar disc, and the base of each
cone upon some surface, or other extent of space; and all
these so equally and compactly interlaced and intermingled
with one another as not only to render all distinction
amongst them the most arbitrary affair imaginable; but to
reduce to an infinitesimal extent, if not to nothing, the dark
spaces between the rays. These are the *diverging cones*;

those from each point being commonly called in the old-fashioned way, and often still called, "pencils of Light," their distinguishing characteristic being that their apex is in some point of the solar disc, or other luminary, and their base in each instance upon some distant space or distant object.

It is, however, at the same time manifest that, from the opposite direction, we have, although only upon the same principle of most arbitrary distinction, OTHER CONES, formed by these very same threads of light, each of which other cones has its apex in the illuminated *point* of space or of surface, and its base on the *entire* area of the solar disc; because from each point of the latter to the illuminated point there extends one of these fine delineations of which all Light is here theoretically and somewhat fantastically supposed to consist. These are the CONVERGING CONES; *i.e.*, converging from all points of the disc to each separate minutest point of the Universe which is presented to it ; and both sets of cones are, we see, essential to the full import and full development of this Radiation Theory, for the sun as well as for all other luminaries. We see that to speak only of a " Convergence Theory," or only of a " Divergence Theory," would be a very partial and imperfect expression of the familiar facts upon which the ideal theory of Radiation professes to be founded; for the whole disc, it is clear, or other luminous body, radiates TO each spot of space or surface presented to it, quite as truly as it radiates AWAY FROM all such spots. Even the theory itself involves no denial of this very obvious fact. The converging cones and the diverging cones are merely different imaginary groupings of precisely the same arbitrarily imagined threads of Light, according as we choose to place our apex in the illuminating point or in the point illuminated. It is unnecessary to add that the number of luminous lines terminating at each point of disc or object is thus in this theory absolutely infinite, in the sense of " innumerable " or indefinite, depending entirely upon our own choice or power of imagination, and not in the smallest degree upon the distance between the object and the disc, or upon the size of either.

Such is the whole Theory of Solar Radiation and of all luminous radiation in its natural and uncurtailed form; but we see at once that we have here no diminution of Light whatever, any more than if Light proceeded like gravitation in one sheet or flood. Every *point* at the confines of the system has here still its full cone of rays with the base resting on the *whole* disc. This also these theorists acknowledge. The more obstinate, therefore, are reduced to the necessity of modifying it. Now, how do they proceed? What modification will it be supposed they suggest, in order to diminish the force of the *whole* solar beam (the converging cone) upon each square inch at the confines of the system.

For this purpose they simply begin by abolishing the converging cones altogether. They adopt what they seem disposed to call the Spoke Theory, in which the rays are, they say, as the spokes in a wheel (Appendix, Nos. 1 and 3). They suppose an arrangement of the rays (but gratuitously, nay, frivolously suppose it) according to which no point of space, far or near, is illuminated by the whole disc (*i.e.*, by all the rays of all the disc) even when perfectly exposed to the whole of it. No point at all of space or surface therefore in this new theory has the cone of rays which naturally belongs to it from its complete exposure to the disc or other luminary, viz., the converging cone; and consequently, very *few* points of space or surface anywhere have even a single ray, *i.e.*, very few compared to the universality of points exposed to the disc. Let this be attentively considered. A single ray, here and there, out of the whole disc (where even so much as that is allowed) is considered a natural and ample supply for every purpose of illumination; the result of which is that there is, as they moreover frankly tell us, in most parts of the system only one such solitary ray, where billions of billions of points have none at all; spotting thus or dotting with spots of Light the more distant portions of space and all the objects occupying these portions.

This form of the Radiation Theory, which is at once both unnatural and arbitrary in the extreme, is more truly as I have said a radiation of dark cones, a Divergence Theory or Spoke Theory, than simply one of Radiation, and is the one usually adopted. Its chief propositions are:—that the number of rays proceeding from each point of the solar disc, or other source of Light, is supposed to be exceedingly limited, and that there are unilluminated cones of much greater dimensions between them; that rays or spokes of Light, as the theorists themselves so aptly call them, do indeed proceed from each point of the disc, but diverging considerably from one another; not therefore by any means proceeding to *all* the objects or *all* the points of all the objects exposed to the disc, but only to some of these, however completely *all* these points and objects may be exposed to the disc; and this without any reason assigned or assignable for the unequal distribution;—that there are therefore dark spaces between the supposed diverging rays (one such space between every pair of rays), which dark spaces, gradually expanding between the rays, more and more, as the distance from the sun, or other source, increases (and this in the true ratio of the pyramid and its square of distance), are supposed to receive into their darkness such objects and such portions of objects as are not in the straight path of these isolated lines or rays; so that even very small objects at a vast distance from the sun, but lying in the path of the ray, are brilliantly illuminated, while at the very same moment, larger objects, nearer the sun, are in utter darkness, from the fact of their lying *off* the lines of light, and exactly *in* the black spaces between the paths of these rays.

Such is this most imperfect and arbitrary form of the Ray hypothesis or Radiation Theory of Light now in vogue; —the only form of it which would at all, even by the aid of an equivocation, enable us to speak of a decrease of Light by distance independently of all medium and all absorption; but it does this, we see, only in the most ridiculously equivocal and inaccurate sense; for the mere introduction of unilluminated spaces between the rays, which spaces are continually becoming wider and wider, does not, in the smallest degree, diminish the number of the rays, nor the luminous force of each ray at any distance whatever from the sun. The whole light of the sun remains under this theory perfectly undiminished at the confines of the system. There is not here the slightest diminution of the solar light by distance on any point to which it reaches. Imperfect, however, and unnatural and arbitrary, and, let me add, unscientific in the extreme, as this restricted form of the Radiation hypothesis is, it is nevertheless this grotesque form of it which all professors and many scientific men understand by that name, and accept as the origin and explanation of their theory for the diminution of Light by that enlargement of the space illuminated, which is vulgarly described among them as " the square of the distance." No professor of Physics disputes this theory about " spokes of Light," however much he may seek to conceal it,—the theory by which all rays diverge steadily and considerably from one another, and (having from the first a dark space between them), thus constitute a Radiation of dark cones, and render every source of Light still more truly a source of Darkness. Instead of grasping the whole of the Lineal hypothesis in its full extent (which hypothesis, however, I repeat, has as little foundation in fact as a lineal hypothesis for Gravitation or for air would have), these scientific men content themselves with one-half of it, or rather with a very much smaller part than one-half of it. Of the ray theory or Radiation hypothesis for Light, which naturally consists of the two parts,—the Convergence theory and the Divergence theory,—they not only take away entirely the Convergence theory, leaving nothing except the Diverging cones; but they also at the same time, and by this very hypothesis, thus added, of non-convergence and non-universality, remove from these Diverging cones, by far the larger portion of the rays belonging to them,—by far the larger portion of each Diverging cone. They suppose the number of luminous lines in each diverging cone (the "pencil of rays"), to be so much reduced as to admit of these lines having gradually widening, unilluminated spaces between every two of them. Instead of recognizing the obvious fact that each object, and each square inch of each object, is illuminated by the whole source of Light, when exposed to the whole of it, they consider that, however completely it may be exposed to the whole source, it receives Light nevertheless from a very small portion of the whole. Instead of seeing that, as above remarked, the hypothesis in question, when complete, requires a thread of Light, or ray, to proceed from each point in the sun's disc to each point of every surface exposed to it, so that there should be, *in the case of each object illuminated,* a cone of light with its apex on each point of the object, and its base on the whole disc, which is the converging cone, as well as a corresponding cone (the ordinary "pencil"), with its apex on some point of the disc and its base upon every portion of every space or surface (which universality of the rays or lines is rendered essential by the natural facts upon which the hypothesis professes to be founded),—instead of recognizing all this, these scientific writers and lecturers,— many of them men of considerable shrewdness in other matters,—suppose, without however assigning the slightest grounds for their opinion, that the number of rays proceeding from each point of the disc and constituting each so-called "pencil" or diverging cone is, *in all probability,* extremely limited; that rays *probably* go from the sun to *some* parts of exposed surfaces, and to *some* parts of exposed space, but not to others; that there is not, proceeding from each point of the solar disc, a separate and special ray or line of light, for each point in the object illuminated, which rays, all thus concentrated at that minute point, however distant, would affect a plant or a retina as much as if it were quite near the disc,—that all this universality has been a mistake; —that on the contrary, there are, comparatively speaking, *very few* of these lines at all that emanate from *the same*

spot of disc, and accordingly a vast number of surfaces and
points of surfaces,—nay, by far the larger portions of our
whole solar system,—which receive none of them; that the
rays therefore do not leave each point of the disc at every
conceivable angle, as one might suppose they would, to
reach every conceivable point of exposed surface or exposed
space, but only at a few, not very well defined, angles in
each "pencil;" and that these rays fall therefore more or less
thickly upon some objects and miss others altogether, just
according to the size of the object and to the direction thus
arbitrarily given to these rays; small objects even when
near the sun, as well as large ones at a sufficient distance
from it, being thus frequently unable to receive any of
these extremely divergent rays at all upon them; while
not unfrequently the larger object near the sun may be in
one of the blank intervals, while the smaller object at a
vast distance is strongly illuminated. And it is most important here to remember that these theorists assign no
natural cause or other reason of any kind for the extraordinary divergence and reduction they thus make and
represent as so probably true, in the lines of light, proceeding from each point of the disc; and not only do they
assign no natural cause for this divergence and reduction;
they assign no reason of any kind whatever for either.
Nor do they indicate any principle that should determine to
which surfaces or points of surfaces rays should proceed
and to which not; the sole requirement being that they
shall not proceed to *all*, nor even to many, however much
all may be exposed to all the points of the disc; and they
regard this hypothesis of theirs (this hypothesis of mere
divergence or Partial Radiation) as what is most probably
the arrangement in nature.

Now, it is quite true and quite clear that in such a
theory,—in this arbitrary and unnatural form of a theory
in itself so arbitrary and so unnatural,—the mere distance
alone without any medium would, in this Radiation of unilluminated cones, determine whether a point of space or of

some surface should or should not be illuminated,—should or should not receive a line of light upon it; but this would no longer be a diminution of light in any received sense of the expression, but merely a diminution of the space or objects that can be illuminated. There would here be no diminution of force (or degree) in any given ray of light, nor in the number of such rays. The whole solar light at any conceivable distance from the Sun, would, even according to this theory, remain undiminished in the slightest degree, and under every circumstance. All that would hereby be diminished, would be the number of points or objects exposed to the source, and from which it could be seen by an eye placed there; obviously presenting a mere equivocation entirely unconscious, no doubt, on the part of these theorists;—also a very transparent, as well as a very useless one. For this theory really amounts to nothing more than a mere confession that certain objects, and certain points of certain objects, are arbitrarily supposed by our theorists *not* to be exposed to the solar disc in their theory, while they allow themselves, nevertheless, to speak and to calculate in the lecture-room as if all such points and objects were fully exposed to it. We are here, however, as in all things, bound to stand by Common Sense and honest Language; and Common Sense, as well as Science, revolts against such conduct and against such language as this. Common Sense teaches us that, if these points and objects are fully exposed to the whole disc (*i.e.*, to each point of it) they receive upon each of them,—upon each point of each,—the whole light of the disc, whether they happen to be in some particular theoretical path or not; and science stoutly confirms this teaching. If, on the other hand, we are bold enough to deny this fact respecting the Light on each object exposed to the Source, we must be bold enough and honest enough to say we do so.

We see then that the Ray Hypothesis for Light gives no diminution of Light at all, nor any room for even an equivocation respecting this diminution, unless we have

recourse to that partial and utterly unnatural form of the
hypothesis, which may be called the Spoke-Theory or
Dark-cone-Theory, the Theory of Divergence, and which is
so generally though timidly adopted, as the correct form
of this hypothesis, by those who lecture and write upon
the subject; and we see without any difficulty that what
this Spoke-Theory affords is a mere equivocation; in which
we say that Light is diminished by distance, when we only
mean that, by this theory, there are, at a distance, fewer
objects and spaces so placed that they can receive any of it.
We see that the alleged fact itself of these dark intervals—
these dark cones—between the rays, even if it were true,
would give no grounds for saying that Light itself undergoes any diminution whatever throughout space, *either in
the number of its rays or in their intensity,* — either
through dilution or any other enfeeblement.

I have dwelt the longer upon this Radiation Theory
(Radiation we thus see of Darkness as well as of Light),
whereby the diminution of Light, independent of all medium
and all absorption, is supposed possible, because to explain
it in all its curious bearings is to show its gratuitous con-
jectures, and so, in the most effectual manner, to exhibit its
unscientific character and to refute it.   If this Radiation of
Light existed at all, it would be universal.   There would be
converging cones as well as diverging ones, and there
would be no unilluminated spaces gradually increasing and
widening between the rays; and if so,—if universal,—this
Ray-Theory would not alter, at any distance whatever from
the sun, the proportion between the solar light and the
space illuminated.   Even if the Theory were the partial
thing supposed,—the Dark-cone-Theory, Spoke-Theory, or
Divergence-Theory,—leaving thus an enormous amount of
the solar system in utter darkness,—much more than one-
half of it, and as much more as the expanding dark cone is
wider than the unexpanding ray,—this would clearly not
give rise to the slightest diminution of the solar light,
wherever the solar Light existed.   This would involve no

dilution, no enfeeblement. The number and intensity of the solar rays would remain entirely unaffected by this circumstance. Even these theorists themselves acknowledge that this is so. The only effect of such a doctrine would be to make it very evident that where there was none of the Solar Light possible there could not well be any diminution of it; that most objects, spaces, points, and surfaces in Nature, even near the Sun, as well as those at a distance from him, which are supposed presented to the disc, receive no Light whatever from it, being really not presented to it at all, the disc being invisible from them,—*i.e.*, from their position in one of the dark cones,—while other objects at their side, and even still more remote than they are, are brilliantly illuminated; and that all the surfaces, upon which any rays do fall, are merely dark spaces *spotted with Light*,—with the spots at greater or lesser distances from one another according as the object is farther from or nearer to the Sun.

Thus far enough has been said to show the equivocal and utterly false sense in which the Spoke-Theory can be said to diminish Light. We have now, however, to proceed another step in this very unscientific land of Theories, in order to point out the supposed connection between the celebrated " Inverse Squares" and this Theory of Divergence, or Unilluminated Cones,—the precise theatre of confusion, in fact, in which the Theory of the Inverse Squares will seem to many to have wholly originated.

Absurd and frivolous as must appear, to all unbiassed and enlightened minds, this Theory of unilluminated *intervals* between the solar rays, which intervals increase as the square of the distance does, and of dark surfaces with strong luminous spots wherever there are any surfaces anywhere lying across the paths of these isolated and diverging rays, it is nevertheless upon this preposterous hypothesis that the further Theory, whose connection with it we must next attend to,—the Theory of the Inverse Areas or Inverse Squares of distance,—is constructed.

This Theory of a perfectly dark cone or pyramid of space along the whole side of every ray of Light, and thus constituting a pyramid of darkness between every two of these rays, treats only, as we have seen, of the diverging rays which form these cones.   It altogether ignores the existence of the converging rays, which have been explained in a former page, as the more important element in the hypothesis of Radiation; and ignores it obviously because it presents no dark cone whatever,—no machinery therefore for this alleged diminution of Light without a Medium; which is here looked upon as the great scientific *desideratum*.   Our business now therefore is with these diverging rays only, and their cones of *utter darkness*, in order to show in what way these helped to originate the Theory of the Inverse Squares. (*See* Appendix, Nos. 10, 12, 13.)

It is important to attend to the fact that we have here no cone or pyramid of Light whatever. The only cone we have is a cone of darkness. The assertion to the contrary is a mere misuse of language. The Cone of Darkness is, in this Spoke-Theory, bounded, it is true, by Light, but only by a single fine line of Light,—not by a Cone of Light. If we had a cone of light, there could not be the decrease of Light which the theorists advocate. No ray, in their theory, has the form of a cone. The ray is held by them to be an infinitesimally fine thread,—in no sense a pyramid or cone. The only cone or pyramid that we have to speak of, in this theory, is the pyramid of darkness which the theory supposes between every two rays, or the compound pyramid composed of many single ones, and placed between any two rays that we may think proper to select as the boundaries of this compound pyramid. It is the fashion, among eloquent lecturers, to speak of the " pencil of rays " as a pyramid or cone of Light.   But this, except in a medium, it manifestly is not.   If it consists, as they tell us, of threads separated from one another by a dark space, and diverging from one another so that this dark space increases in width as they extend, it is, in that case, distinctly a pyramid or cone of Darkness, not of

Light. Or if there be several such cones placed beside each other, they thus constitute one large cone of Darkness with threads of Light in it, and spots of Light upon the black surface at the end of it. It is still a black cone or pyramid,—a combination of several fine cones of Darkness which are united into one large one, and merely separated from one another by these fine threads; and it is only the dark unilluminated intervals (between diverging rays), whether these intervals be compound or simple, which have the form called "the pyramid" or "cone," and which can have the calculations belonging to such a form. No one pretends that a ray has the form of a cone or pyramid; and the cone of Darkness cannot be called a cone of Light merely because it has rays of Light around it, or even some isolated threads that extend along the cone, inside it.

We have seen in PART I. of this Treatise that the areas or sections of the pyramid or cone, in the diagrams of Geometry, are in the same ratio as the squares of their distance from the apex; and it has been there explained that, and how, this is true of all areas in nature,—of all the objects around us,—although these are almost never actually sections of any existing pyramid. Every object has its area or size, and every two areas that differ have, in Geometry, their pyramid or cone; and these areas or sizes differ as the square of their distance from the hypothetical or geometrical apex differs. So that, when we know the relative areas of bodies, we can tell their relative distance, and *vice versâ*, when we know the relative distance we can tell the relative area. This is the universal law of sizes or areas.

We further see that when we have a cone made by and between the divergent rays issuing from a luminous point with this luminous point therefore as the apex of the cone, we have, by this Divergence Theory (or Theory of Unilluminated Cones, commonly called the Radiation Theory), a Radiation of Darkness as well as a Radiation of Light. We have fine threads of Light proceeding from the lumi-

nary, which, in a very remarkable manner, diverge or
separate from one another, as they proceed; so that the
unilluminated interval between every two of these isolated
and diverging rays becomes, by this curious theory, wider
and wider, as the delineating rays extend, these intervals
being dark cones or pyramids of space, with their dark
sections or areas, wherever we choose to mark these off,
and upon which not even one spot of Light appears; for
the spots of Light can only be upon the sections of these
cones when these cones are compound.

The distinction here alluded to requires attention. The
unilluminated cones in question are either simple or compound. The cone lying between any two neighbouring
rays is the simple one, no two such rays are without their
dark cone; and several such cones together constitute
the compound cone. We easily see that the simple (or
single) cone has no rays passing along inside it at all;—no
spots of Light, therefore, at the end of it, i.e., upon surfaces at its sections. We easily see also that the compound
cone has as many of these rays, running within its boundaries, from the apex to the sections, as the number of the
single cones which go to constitute it. We thus see that
the compound cone has several spots of Light, thrown on
surfaces at its sections, by this theory, while the uncompounded one has none at all. The only two further points
of which it is necessary to remind the reader are: (1) that
the rays, in each compound cone of darkness, thus formed
by rays issuing from a point, neither increase nor decrease
either in force or number through their whole course within
the cone from the apex to the section, and throw therefore
the same number of luminous spots upon the larger (i.e., the
remoter) areas or sections of each cone as upon its smaller
sections,—neither more nor less; and (2) that as in this
theory there is no medium to scatter or diffuse the light of
the ray, from side to side or up and down, it is only as a
spot that its Light can appear upon the area or section that
receives it. We see, then, how it happens that, although

the simple cone has its sections and objects, whether large
or small, near or remote, entirely unilluminated, the compound cones have their sections and objects spotted with
Light, but having, between these spots, all the dark spaces
of the uncompounded sections, and these spaces becoming
wider as the sections become larger. We see here also (3)
how it happens that mere points at vast distances from
the sun are brilliantly lighted, according to this theory,
from merely lying in the path of some ray, while even
objects of some size near the sun are wholly unilluminated,
because they lie *off* the path of any ray whatever, and
entirely in the path of a dark cone.

What we have now further to attend to is that these
unilluminated cones or pyramids, whether simple or compound, increase rapidly in *size* while, in the language of the
theory, the rays which are connected with them, decrease
as rapidly in *number ;*—nay, that the rays decrease thus, in
their way, precisely *because* the cones increase in theirs;
and *at the same rate* as the cones increase. It is indispensable that we should understand this part, and this language,
of the Theory.   These black cones (and there are here no
cones except black ones) become wider as they become longer
and longer; but the rays which form them, and between
which they exist,—also the rays which run, inside, divergingly along the compound cones,—do not, as they become
longer, become either more powerful or more numerous.
Neither do they become less so.   What these rays were in
the beginning of the pyramid or cone they are at the end.
They remain unchanged ; the same in number and the same
in force.   Whether it be simple or compound, it is the dark
cone or pyramid alone which changes. The rays do not. But
we are told that, since these rays remained unchanged, they
diminish ! and diminish, moreover, as fast as the cones
themselves and their sections become enlarged !   This is
the language of the Theory; and this point is what we
have now distinctly to disentangle and to grasp.

Let us attend closely to the alleged facts.   When it is

said that, at the same rate as the sections of the dark cone
increase in *size*, the delineating rays, between which it exists,
as also the rays within it when it is a compound one, decrease in *number*, this only means that they do so, *comparatively speaking*. They do not really decrease. It is merely
self-deception when we think or say they do. They only
become fewer and fewer, when compared to the larger
sections on which they fall in the cone they traverse or
delineate, than they are when compared to the smaller
sections of it. It is only in this inaccurate and frivolous
sense of things compared, that the rays are said to decrease
in force and number when the distance increases although
they remain as they were at first. Yet such is the technical
language here adopted by the Theorists. It only means (I
repeat) that, comparatively speaking, the rays, although the
same in number, *seem* fewer and weaker when the areas
illuminated are large than when these are small. Comparatively speaking, they are fewer and weaker, but not really
so, as the professors teach. When the sections of the
cone increase in size (*i.e.*, at the greater distance in the
cone), the light falling on them,—the number of rays falling
on them,—*seems* to decrease, but does not really decrease,
for by the theory it remains the same. We thus again
return to the principle illustrated at pp. 62 and 67 with the
bread and butter and which we may here illustrate with
bread and jam. If we give to half-a-dozen children precisely the same amount of bread each, and to each a spoonful of jam to eat with his bread, by then giving one of them
a second spoonful of jam, we, in this phraseology, reduce
his bread to one-half of what the rest have, and must give
him twice as much bread as to each of the others, in order
to give him as much as each of them has! In this way
our Lecturers assure us that the larger area has a lower
degree of light on every part of it than the smaller area
has, even when each area has precisely the same degree of
it; and this merely because it is the larger;—that where
one area is double of the other, its light is half; where

treble, one-third, &c.; and this is what their theory means
when it says that the light is diminished by the distance,
for it is distance in the cone which gives the larger area.
We see here also what a misconception it is to say that the
same light which is on the greater and the smaller area
becomes "diluted" on the larger one. Nothing of the kind
occurs. All that their theory authorizes them to say is that
this same degree of light on both areas is disproportionate
upon the larger one; and that it would be necessary to
increase this light upon the larger area if we wish to make
it proportionate to that upon the smaller one. But they do
not use the word "proportion." They simply say that if
one of two areas is only half the other, the larger area must
have twice the degree of light upon it which the smaller
has, in order to have the same degree of it as the
smaller !

Such is the equivocation and merely relative sense in
which the "Spoke-Theory" and its professors teach that
Light is diminished by the divergence of its rays. We have
now to see in what ratio of the distance this relative
diminution takes place (Appendix, Nos. 12 and 13); and this
is easily measured for us by the dark spaces or cones
between the rays. Since the ratio in which the sections of
these dark cones increase is the square of their distance in
the cone, it is in this ratio also that the accompanying rays,
whether traversing, divergingly, the cones or delineating
them (i.e., whether inside or outside the cones), have this
appearance of decreasing. These rays which thus accom-
pany the dark pyramid of space that lies between them and
around them become proportionably fewer, and therefore
proportionably weaker, the theorists tell us, because the
sections which they have to spot with light become larger,
and, moreover, at the same rate as these spotted sections
become larger. The sections become really larger. The
rays become only relatively fewer; but they undergo this
relative or seeming diminution at the same rate as the sections or sizes of objects really enlarge. This is because the

sections change in size, while the rays always remain unchanged in force and number. Since these sections, therefore, really enlarge as the square of their distance in the cone becomes greater, the rays become relatively fewer and therefore relatively feebler in this same ratio; viz., as the square of their distance in the cone becomes greater. For to diminish, in any sense of the word " diminish," as the area enlarges, is to diminish as the square of the distance in the cone increases.

To recapitulate now what we have here gone over, we see that, throughout the cone of black space thus marked out by rays, the dark areas or objects presented to the apex increase directly and absolutely as the square of their distance in the cone; we see also that each spotted area,—each spotted section or object,—in the compound cone, however near or distant, however great or small, this area may be, has upon it, by the theory, precisely the same number of rays or spots of Light, and each spot or ray in the same intensity, force, amount, or degree of Light as when these rays first left the source; we see, therefore, that the rays being thus, by the theory, fewer comparatively, but only comparatively, on each greater area, and more numerous comparatively and in appearance, but comparatively and in appearance only, on each smaller one,—these rays *seem* to increase and decrease inversely as the areas do on which they fall; *i.e.*, in a ratio inverse to the square of the distance. They increase and decrease thus inversely and relatively while the areas or sections themselves of the cone increase and decrease directly and even absolutely in the same ratio. We see that, although these dark cones do, by this theory, really increase in size, between and around the rays by which they are delineated or, divergingly, lengthwise traversed, yet the rays themselves do not increase in number. They remain in all respects unchanged by the distance; and for this reason are said by the professors of Physics to decrease! nay, to decrease as fast as the dark areas, which they speckle with light, increase !

We may here profit by this long but indispensable digression on the "Spoke-Theory" and its connection with the law of the Inverse Squares, to illustrate three points connected with this law and already adverted to in explaining the theory of it, yet too often lost sight of even by the theorists themselves.

1. We see that when the professors consider the light upon the smaller and larger area as always equally strong when equally exposed to the same source, but as less, on each "unity of surface," upon the larger area than it is upon the smaller, they only mean, and can only mean, that it is comparatively less; that the larger area ought naturally to have the stronger light; and that in order to bring the two unequal areas to an equality of illumination, it would be necessary to increase the strength of the light at each point of the larger area by as much as the larger area is greater than the smaller. They forget that this could not be done without reversing the dark cones of their theory, placing the base of these cones instead of the apex on the source, and increasing the spots of light in proportion to the increase of the area instead of leaving them unchanged. In short, they hold that the larger regions of space should have this higher degree of illumination than the more contracted regions, merely to keep them illuminated upon a par with these more contracted regions, and that, as this is not the case, the light of the sun is (comparatively) diminished through the enlargement of the regions or other areas illuminated by it. They here, in their statements, omit the word "comparatively;" the importance of which word, in such a case, every one can judge of.

2. Nothing can be clearer than it is made by this Spoke-Theory, commonly called the Radiation-Theory, that it is the Enlargement of the Area on which the rays form spots and which stands at the greater distance in the cone,—not at all a diminution in the number and intensity of the rays which fall upon that greater expanse,—that makes the

rays which fall upon it less *in the only sense in which they are less;* viz., less in proportion to the increased size of the object or area which receives them in the more and more isolated manner required by the theory.

3. Another point which our analysis of the Spoke-Theory may help us to discern more distinctly is, that it is distance in a cone or pyramid, *i.e.*, distance from an apex or angle, —not, as so many professional physicists have imagined, distance from a luminary (window, sun, or lamp), that makes the illuminated objects (the areas) greater or smaller. What seems to have led to confusion on this point is, that some writers, instead of using any other delineation of a cone, have illustrated the geometrical law of areas with the cone of darkness formed by rays issuing, at an angle, from a luminous point, and thus bounding the dark cone which by the theory is formed between them. (Appendix, Nos. 12 and 13). We see clearly, however, that it is not the luminary nor the distance from it which here makes one area or section bigger than another. What does this is the angle at the apex (whether formed by rays or not), and the increasing distance from each other, maintained through their whole course by the lines which form this angle, and which therefore at each point require a certain object, or area, to fill the space between them. It is thus that the objects or areas which subtend or are opposite to this angle are greater or smaller, according as they are farther from or nearer to it ; not at all according to their distance from some luminary. We see also that the only objects whose size is regulated by the angle in question and its sides, are the areas which subtend this angle. Other objects, lying within the cone or extending beyond its sides, would not be made greater or smaller by this angle nor by their distance from it; *a fortiori* then, not by their distance from a window or other luminary. Nor could this angle make such objects have any relation of size or distance to one another. The planets, for instance, are not made greater or smaller by their distance from the sun, nor the pictures on the wall by their

distance from the window. Yet all such objects have the size belonging to a particular distance from an apex, and the distance belonging to a particular size. In a word, nothing can be clearer than it is that the " distance " mentioned in this law and theory of the Inverse Squares is the *distance which defines the size* of an object,—distance from the Source of Size (pp. 23, 56); and nothing clearer than it is that this defining distance is not at all, as so many professional Physicists think it is, the *distance from a lamp or other source of light.*

Enough has been now said upon this subject for our present purpose; which is merely to show how the old Theory of Divergence or Unilluminated Cones and spotted Sections (the Spoke-Theory) gave rise to the clumsy assertion that the spots of Light upon the dark sections of the compound cones are, although in number always the same, in the same cone, yet few or many, according to the size of the dark spaces they are upon,—few when these are large, and many when these are small ; or, in other words, that the illumination of objects, areas, or regions is in a ratio inverse to the size of the objects, areas, or regions illuminated ; and this, although even by this Theory of Radiation, there falls always the same amount of Light,— the same number of rays and each ray in the same intensity,—upon each object equally exposed to the same source. Enough also has been said to explain from this Theory, that when the size of an object is said to be as " the square of the distance," the term " distance " here only means, as every Geometrician knows, distance in the cone, pyramid, or other diagram, which is the only distance that makes one object greater than another, not as professional Physicists suppose, distance from a lamp, or window, or other source of Light,—a sort of distance which has no such effect as this of defining the size or area of the object upon which the rays fall as luminous *maculæ* upon a black ground.

FIFTH SECTION.

RECAPITULATION OF THE RECEIVED THEORY RESPECTING THE
DIMINUTION OF LIGHT BY ENLARGEMENT OF THE SURFACE
ILLUMINATED; AND OF THE PHYSICAL IMPOSSIBILITIES
CONTAINED IN IT.

FROM what has been said thus far, we see that the common
theory—the theory of the Inverse Squares—consists mainly
of the following four elements, which cannot be too much
kept before people's eyes, and must never be lost sight of
for a moment by those who desire to understand the
question :—

1. The diminution of Light now under consideration is not
supposed to be caused at all by the distance itself which
there may be between any space or surface and the sun or
other source of Light (for diminution by distance is what
occurs solely in the case of a medium), but merely by any
enlarged space presented transversely to the sun or other
luminary; and this enlarged space, if not otherwise defined,
may be described as that which would geometrically result
from its greater distance in the cone, *i.e.*, from the
greater distance existing between it and the apex in
its appropriate pyramid, and would be in the ratio of
that distance squared, as has been already explained in
PART I. (*see* p. 18). It is most important to stand clear of
the flagrant equivocation here employed, apparently in utter
unconsciousness, by many of our theorists (Appendix, Nos. 2,
3, 5, 15, &c.) when they speak of Light and its diminution,
in their Theory, as dependent upon distance from a *source*,
for this is not really what the more enlightened of them
mean; nor as has been seen, would this at all suit their Theory.
This diminution of Light by distance from the source is
what happens only when Light passes through a medium;
and this is then a result of the medium's density as well as
of its length, or the distance to which it extends. They
are very far from holding that distance from the source,

without a medium, diminishes the Light. They admit, on the contrary, that the increased length of the sun's rays greatly increases the illumination of space, as far as that length alone, or mere distance from the sun, is concerned. They fully admit that it is only the enlarged transverse space, area, or surface, itself exposed to the disc, and not its distance from the source, that produces the diminution which their theory treats of, the enlarged space spread out on every side around the sun, not the enlarged space lying between the sun and some point of that transverse space; for, I repeat, they completely recognize that in that (the third) dimension of space the solar illumination is, on the contrary, enormously increased by the increase of distance, holding, as they seem to do, that a long ray gives more light than a short one.

2. These theorists further hold that all rays, whether we suppose emission-rays or wave-rays, lie in straight lines, from the sun to the remotest limits of the system, completely undiminished in their whole length, *i.e.*, losing nothing of their force nor of their number, nor as most professors admit, of their proximity to, or rather lateral continuity with, one another throughout all that distance,—having no spaces whatever between them. These writers do not here pretend that there is the least diminution of Light *by distance*, in their theory, but quite the reverse, in consequence of the sun's rays being all thus prolonged in unabated force to the confines of the system, increasing thus in luminous masses, rapidly throughout the whole of space. This perfect equality in the sun's light, *i.e.*, in the force and number of the sun's rays, *at every point of space*, from the disc outwards, is a very important item of this theory, and fully recognized as an important one by all who hold the theory.

3. They admit moreover,—at least, most of them nowadays insist strongly that from *each point* of the disc to *each point* of the most distant space, there is one of these unimpaired lines of light,—infinite numbers of

them, therefore, proceeding from *one and the same* point of the disc to the whole of the space confronting it, and also from the whole of the disc to every separate point of this confronting space however distant; thus forming *cones* of rays without number,—some converging, some diverging; some whose base is on the whole disc, and their apex on some point of distant space; these are the converging cones; others whose apex is on some point of the solar disc, and their base on some extent of distant area; these are the diverging cones; and the Theory insists strongly that, as above stated, each of these rays, at each point of *its path*,—at each point of each concentric sphere of space it traverses,—retains precisely the same undiminished force with which it left the sun, as well as that, to each minutest point of the most distant space, a converging cone of these rays extends whose base is on the solar disc (*see* p. 90).

4. Then fourthly and finally, combined with the three foregoing principles respecting (1) the enlargement of the space to be lighted; (2) the unimpaired force of the ray throughout its course; and (3) its rectilineal progress outwards, — but unintelligibly combined with them, — the Theory goes on to teach that, in all this enlarged region of transverse space,—in all the space at right angles to this outward progress of the Light,—in all the space or surface presented, and spread out, transversely to the lamp or to the Sun, so as to receive its rays,—the light here *loses its force,* spreads, *i.e.*, deviates in all directions from the straight line, and thus, it is said, becomes attenuated, enfeebled, dimmed, diluted, by its greater subdivision over this enlarged space, although, on every point of it, there falls a cone of converging rays in undiminished force, just as on every point of the contracted space which is nearer to the Sun or to the lamp.

Such is the whole theory. Such all that they mean when they say that Light decreases inversely in the ratio which they self-deceivingly call the Square of the distance,

which, however, we easily see, only means, as all the more
enlightened admit, in the ratio of the more extended space
illuminated.   Whenever there is no increased extension ;—
no enlargement of the space illuminated, there is then,
according to these more enlightened of the professors, no
diminution of Light whatever.   The question for us on this
point is: How can there be any of this diminution, even
where there is this enlargement of the space illuminated?
Upon the least reflection the physical impossibility of this
is obvious.   For how can Light deviate from its straight
line,—its rectilineal course,—in order to spread and diffuse
itself, where there is no medium?   It cannot.   Or where is
the redistribution of Light resulting from the enlargement,
when the source is central, or *one for all?*   There is none.
This could only result from some subdivision and distribution of the source.   And, which is the portion of the transverse space exposed, upon which the *whole* experienced
light of the disc is not to be found?   Which is the point of
it upon which, according to this theory, a ray from each
point of the disc does not fall undiminished, brought there
in its own converging cone from the disc?   Clearly there
is none.

If the solar light proceeds, as they say it does, unchanged in intensity to each point of space,—the remotest
as well as the nearest,—what is the meaning of saying
that there are some points of the transverse space to which
the solar light does not come directly from the disc but by
spreading without a medium,—points upon which the sun's
light is not in this undiminished force which the theory
teaches, but either not upon them at all, or considerably
weakened and diluted from there being more points to be
illuminated than the rays can come to in straight lines?
Are these words supposed to have any coherent meaning?
If there were some points of transverse space to which,
according to the theory, the converging cones of rays could
not come, or did not come, on account of some overpowering
divergence, as in the Spoke-Theory of some professors

(*see* Appendix, Nos. 1 and 3); or if these converging cones lost some of their rays, or the rays some of their force, in consequence of the distance they had to travel, the thing would be intelligible enough. But the theory denies *in toto*, and most justly, that any such obstacles exist to an equality of illumination; and the theorists profess not to see, in this part of their hypothesis, its incompatibility with every known principle of nature and of common sense.

It is clear then, from what has been said, that we have no less than three separate cases of flagrant physical impossibility crowded together in this one Theory.

1. *Two different degrees of Light supposed to be on the same area from the same source.*—It is alleged that the same degree of light upon any area differs according as this area stands alone or forms part of another area,—on the square foot, for instance, when it is alone, and on this same square foot when, in conjunction with others, it forms a square yard. (*See* p. 66 and p. 78.)

2. *Light supposed to expand or leave its straight line, and spread if we give it room.*—The unexpansive nature of light, —the straight line theory,—renders it impossible for light outside all medium to *stretch* or *spread* out of the straight line, as air does, without the spreading of the source; *i.e.*, to exist at any point which it cannot or does not reach *directly* in perfectly *straight* lines, from the source. (*See* pp. 21, 60, 62, 63.)

3. *No space here existing for Light to expand in, even if its nature admitted of its doing so.*—Even if it were true that Light was of an expanding (or deviating) instead of an un-expanding nature, yet, as each point of all transverse space is fully and equally occupied, *i.e.*, receives the whole Light, from the source, conveyed to it in the cone terminating at that point, there are no points,—no spaces,—left unillumi-nated, into which Light could by any imaginable possibility spread and expand from other points and spaces better lighted than they, and thus become diluted in consequence of this expansion. As stated, however, in the last para-

graph, a straight line does not expand; so that even if there existed such points and spaces, no light could reach them.

SIXTH SECTION.

TWO EFFORTS MADE TO JUSTIFY THE THEORY OF THE INVERSE SQUARES; VIZ., THE THEORY THAT, WITHOUT ANY ACTION OF A MEDIUM, THE GREATER AREA ALONE PRODUCES A DIMINUTION OF LIGHT,—THAT, INDEPENDENT OF ALL ABSORPTION, THE LIGHT DIMINISHES UPON THE GREATER OBJECT OR AREA, IN PROPORTION AS IT IS GREATER; *i.e.*, AS ITS SQUARE OF DISTANCE IN ITS PYRAMID INCREASES.

OF the manifest incongruities with nature's laws pointed out in the foregoing section, these lecturers seek to offer no explanation. Is it that they see they can offer none? or is it that they do not see the incongruities? Be that as it may, they never even advert to them, although they state and restate, as fact, the hypothesis involving them, in every book and every lecture, and, what must astonish us as much as anything else, are listened to.

But they not only take no notice of the physical impossibilities so patent in what they teach; they become immersed in all sorts of further incongruities in their efforts to justify and prop up their hypothesis. They do not attempt to answer the arguments which common sense as well as common logic brings against their notion that, independent of all medium, Light is diluted by distance from the source (which is the grossest form of this misconception) or even "thinned out" and rendered feebler by being spread over a larger surface, or diminished in the figurative sense of "proportionate magnitude,"—in the figurative sense of there being less of the force and intensity of the Light at each point of the greater area, merely because it is the greater, although there is, they admit, on the greater as well as the smaller area precisely the same amount of this intensity, coming direct to each point from the sun or other central source. They do not seek to explain or answer any of these points. Nay, overlooking all such matters, these theorists, as may be seen in the extracts from their writings given in the Appendix, insist, with remarkable *naïveté*, upon their incoherent statements; and just as if no such incoherencies existed, they merely employ themselves in forming hypotheses and contriving illustrations whereby their theory may be made plausible and natural, they think, to "unscientific people," and even probably be justified in their own eyes. I shall do no more here than mention the two more remarkable instances of this,—viz., the Argument from Perspective and the Argument from Experiments; in both of which Arguments what is effected by the Medium is held to be identical with that which is effected without it; while, in the Argument from Perspective, the confusion between Physical and Physiological Optics becomes perfectly astounding, the diminution of Light by Distance from the eye being, throughout, confounded with that by Distance from the source; and both these Distances also, still further confounded, as usual, with that which is neither; viz., with that which is called "*Distance in the pyramid and cone of geometrical diagrams*,"—a sort of distance, depending wholly on the relative Areas of any two Objects and having nothing whatever special to do either with distance from a source of Light or with distance from an Eye.

In this Argument drawn from the phenomena of Perspective, the Physiological fact that Light is not diminished in degree and brilliancy upon a self-luminous area by its Distance from the eye, notwithstanding our powerful medium and notwithstanding the Physiological or apparent contraction of the self-illuminated area, is brought to establish the supposed Physical diminution of Light upon an area not self-luminous,—a diminution of Light supposed to result from the mere enlargement of the illuminated area, without any action either of Medium or of Perspective;

while the second Argument, drawn from a system of Experiments with a medium, which we must here call " Professorial Photometry," inasmuch as it is not employed except by Professors, deals only with the area which is not self-luminous,—the Physical diminution of Light which takes place upon this area in consequence of distance from the source, where there is a medium, being represented in this strange Photometry, as being identical, in amount and ratio with the Physical diminution which ALSO takes place upon it, in consequence of the same distance independent of all medium and all absorption. We shall speak of each Argument separately,—that from Perspective and Medium jointly, and that from Medium alone. It will be seen that both Arguments, as far as they prove anything, prove the very contrary of what they are brought to prove.

1.—*The diminutions from Perspective and from Medium employed to prove the law of the Inverse Squares; i.e., to prove that diminution of Light which is supposed to result from nothing else but the enlargement of the space illuminated.*

In our dense Atmosphere, which enters into all our experience and experiments respecting Light, there is always a diminution of Light upon every area, self-luminous or not, when at a distance either from the eye only or from the source only or when, as often happens, at a distance from both; and this diminution, the amount of which depends upon the amount of medium traversed, is consequently, *cæteris paribus*, greater or less according to this distance of the area from the eye or from the source or from both. Of course, however, if there is no area to be taken account of but the source, there is then no distance from the source that requires to be taken account of.

Such is the common fact of our experience in this matter; and to this we must here add another also much insisted on by these writers; viz., that the physiological or *apparent* illumination of the diminished area, *i.e.*, the illumination seen upon the area at a distance from the eye,—is not less intense and strong, although there is less of it,—not less in degree although less spread out,—in consequence of increased distance from the eye. These two facts no one disputes. They belong to daily life and daily experience; and the latter wholly to Physiological Optics.

The Medium diminishes the Light upon the distant area not only physiologically,—not only for the eye,—but also physically; *i.e.*, in its illuminating, chemical, and other physical effects; and the Perspective diminishes the area itself, but for the eye only; *i.e.*, only physiologically. For the eye therefore, or physiologically, the Light existing upon the distant area being apparently drawn more together, would, if undiminished by the medium, increase in intensity in proportion as its extent is thus apparently diminished (just as when a dozen gas-burners are grouped together instead of being dispersed over some large space, or as a lens acts); but, being, at the same time, physically diminished by the Medium, does not thus increase in intensity in proportion as its extent is apparently diminished. The result of common observation is that it remains much the same in intensity for the eye, whatever may be the apparent change in the area; so that the *real* effect of the medium counterbalances the apparent effect of the apparent contraction in the area. The medium causes one diminution,—the diminution of the Light. The Perspective, or distance from the eye, causes the other diminution,—that of the area over which the Light exists. So far all parties appear to be agreed.

Now let us attend to the use which the Professors make of these facts, in their efforts to uphold their theory of the Inverse Areas (or " Inverse Squares," as they prefer to call it), and to the result of their reasonings.

They begin by saying that, in all this, it matters not whether we understand the word " distance" to mean distance from the eye, in Nature, or distance from the source of Light (window, lamp, sun, &c.), also in Nature, or distance in Geometrical diagrams, from the Apex of their pyramid. Distance in any case (they say) is distance, and the square of the distance, in any case, is the square of the distance, without any reference to *what distance* is meant. All their statements involve this error and this confusion; which of course, except themselves, no scientific man can sanction.

They then argue that it matters not, for the diminution of Light, whether a medium is or is not present. They say that, since in every case the self-luminous area is diminished to the eye (*i.e.*, seems diminished) in inverse ratio to the square of its distance from the eye, while, except in extent, the illumination on the area remains undiminished to the eye, the illumination itself must *therefore* be also here diminished, under the Medium, in the same inverse ratio of the distance, in order that this identical *reduction* should have resulted. Light, therefore, in our medium (they argue), diminishing to the eye, in inverse ratio to the square of the distance from the eye, it therefore does so also independently of all medium, and in addition to the absorption which results from the medium, and which is always, we are told, going on in a geometrical progression.

This Argument, drawn from the Perspective, applies of course as much and as clearly to the illumination upon areas which are not self-luminous as to that upon those which are ;—to that upon the lunar disc, for instance, as much as to that upon the solar disc, and to that upon the whole illuminated expanse of the solar system, or (which is the same thing) to the concentric spheres of space around the central sun, with their elastic illumination (pp. 24, 33,), as much as to the elastic illumination to be seen (*i.e.*, the apparent illumination) upon the disc of the sun itself, at different distances from it. But although the argument applies equally to both kinds of areas,—to the area which reflects and to the one which is self-luminous (the only essential condition being that the light spoken of pass through a medium, as we always experience it), it may be well, for the sake of precision and simplicity, to speak here only of self-luminous areas,—of the illumination, for instance, which expands and contracts physiologically,—*i.e.*, to the eye,—upon the solar disc, and upon other surfaces which are self-luminous; and the fact which these writers bring forward is, as we have seen, that the illumination on the area,—not, of course, its illuminating power, but the Light seen on, or in connection with, the area, as long as the Light remains at all visible,—remains the same,—of the same strength and brilliancy, although the area itself decreases in size. They accordingly argue that, since the area decreases under the action of Perspective law, in the ratio called the Inverse Squares, the brightness or illumination must decrease under the action of the medium, in the same ratio precisely; for, otherwise, according to their theory (p. 18), this illumination would be increased as much as the area is decreased instead of remaining, as it does, the same. Light is therefore physiologically diminished, with a medium, as well as physically without one, in a ratio inverse to the square of the distance.

In reply, then, to this Argument from the phenomena of Perspective, it is enough to observe (without referring to its other absurdities) that, as there is always a great diminution of Light going on in a dense medium like our atmosphere (which these writers themselves acknowledge and describe as proceeding in a rapid *geometrical progression*—see p. 15, and Appendix, Nos. 1, 15, 16, 20), the diminished amount of Light seen by us *upon* the solar disc or other luminous body being the result of a diminution in inverse ratio to the square of the distance, would, if true, prove (contrary to what is intended), that this is not the ratio of any diminution ascribed to Light independent of the medium; inasmuch as, when we have deducted from the whole decrease the very considerable portion of it, which is due to Medium, and the existence of which these writers fully recognize, we have a balance entirely different from that supposed to be required by the Inverse Squares; the Argument from

Perspective thus disproving the very ratio of diminution
which it is adduced to prove.

But, besides all this, there is no scientific man, un-
hampered by theories, that does not instantly see the preposterous inconsistency of supposing that the diminution of
Light by a medium can be regarded as identical with anything we may imagine respecting a possible diminution
of Light without one.

2.—*The Experimental Photometry of the Professors, wherein
the diminutions which happen where there is a Medium, are
adduced to prove the existence of those which are supposed to
happen where there is none.*

What the Professors of Physics seek here to establish by
their Experiments (Appendix, 2, 3, 5, 15) is that Light without any action of a Medium,—without its undergoing any
absorption,—becomes diminished in proportion to the increase of the illuminated area ; and this point they profess
to render clear to any ordinary intelligence by the aid of
our dense atmosphere and a few "candle-flames," with a
screen or two and a rod. They proceed to place these
articles in position very much as a conjurer would do, and
then ask us : Do we not see ? Yes, no doubt, those who
can see only what they see with their eyes, see all that the
Lecturer desires, and admire his profound photometrical
experiments. But how is a thoughtful man to see all
this ? What is the scientific man to think of such a
lecture, and of such instruction ?

The object of the Experiments in question is to show
that, independently of all Medium and of all Absorption,
Light diminishes inversely as the Square of the distance.
The Professor undertakes to do this in a medium. With all
the liveliness, and *air dégagé* that a Professor ought to
exhibit,—sometimes even with some effort at histrionic
effect,—he places everything, and shows us that the reductions of Light which take place in a medium, are exactly in

this peculiar ratio,—exactly as the square of the distance in the cone or pyramid of sizes. Although by the professional theories on this subject, there are *two* diminutions going on at once,—one by the Medium and one entirely independent of all Medium,—yet here under the action of the Medium, we have only one result, and this, the experimenters admit, the result known as the diminution commensurate with the square of the distance; the result or action of the Medium being, for the moment, we are taught to believe, completely paralysed by the grander action of the Inverse Squares, or Enlarged Areas; and this, even while no other action appears except that of the Medium. This, it must be admitted, is a peculiar form of experimental Teaching. How can we show what change is caused in Light outside a Medium by that which it undergoes within one? If the four "candle-flames," placed at a particular distance from an area, *in a Medium*, produce a particular amount of Light, how can we offer that as evidence that, *outside all medium*, they would produce the same effect or any effect, outside a medium, at all? Where is the Logic of such an Experiment as this? Where its Common Sense? But it is useless to dwell on such a subject. It can be hoped that a better Era of research is coming,—an Era in which the unpaid man of science will not be so rigidly excluded by the paid ones, from that "fair field and no favour" which some enlightened minds have seen to be so essential to Scientific Progress.

Here then also, as in the case of Perspective, the "Experiments" with a Medium and a screen and a few candles prove precisely the contrary of what they are brought to prove. They show, in the clearest manner, that when due allowance is made for the absorption of our dense atmosphere, there does not remain the amount of diminution here intended to be established, and here required in order to support the theory of the Inverse Squares. These Experimenters all tell us that there is this diminution by a Medium going on in geometrical progression at the same time as their supposed diminution by the square of the distance. But where or when do they take account of this double diminution thus simultaneously going on in all their " experiments?" When proving their Inverse Squares with their Medium and their candles, and the rest of their photometrical machinery, why do they never mention their curious law of geometrical progression, whereby all Light in our atmosphere is always enormously diminished, and can be almost as much so as some suppose it can by the Theory of the Inverse Squares? Why should not the diminution they find upon the screen of their Photometer be the sole result of the medium they are employing? Or, do they know that they are employing a medium? How is it possible not to be astonished that such Experiments should be called " Science," and be taught unchallenged at the Royal Institution and at the Royal Society?

END OF PART II.

PART III.

THE LOGICAL IMPOSSIBILITY OF THE LAW IN OPTICS CALLED
THE LAW OF "THE INVERSE SQUARES."

PART III.

FIRST SECTION.

DIFFERENCE BETWEEN THE THEORY AND THE LAW. TWO
INTERPRETATIONS OF THE LAW VERBALLY POSSIBLE. THE
ORDINARY ONE. USE OF THE TERM " INVERSE."

Difference between the Theory and the Law.—We have so far
examined the THEORY of the Inverse Squares ; and we have
seen that, according to this Theory, it is the larger object,—
the object with the larger area,—which makes the Light
less than the Light on the smaller object, while we know
from other considerations, that it is distance from the apex
which makes one object's area larger than another's, not
distance from the window or any other source of Light.

We now come to speak of the LAW which is generally
supposed to express this Theory, and is therefore called,
also, the Law of " the Inverse Squares." It will be seen,
however, that the Law, in the only meaning ever given to
it, treats of quite a different subject, and is no expression
whatever of the Theory so called. The Theory applies to
the illumination of all areas irrespective of their distance
from the source; whereas in the Law this distance from
the source is everything. The Theory is to the effect that,

when each point of each surface is equally exposed to the source, objects of different areas are illuminated inversely as the areas, *i.e.*, less when these are larger, and more when these are smaller ; and this without any reference whatever to their distance from the lamp, window, sun, or other source of Light; while, as I say, in the Law, *the distance from the source is everything.* This law is that, if the distance which the Light traverses is divided into any number of equal parts, the Light, remaining after the first division of the distance, is always four times greater than that remaining after the second division, and always nine times greater than the Light left after the third, however short these distances may be, and although they are supposed to be entirely free from Medium. The difference then between the Theory and the Law is simply enormous, as the attentive reader will have no difficulty in discerning. Our sole business, however, now is with the Law,—the celebrated Law of the Inverse Squares ;—to point out how it is interpreted, and how utterly it is without any data or any Logic ;—also how, in various experiments, its absurdity admits of being easily shown.

I am aware that to many of those who read these pages, and hear this Law perhaps for the first time distinctly stated, it will seem quite superfluous, especially after the explanations in PART II., that I should enter further upon the discussion of a principle so obviously false and frivolous ; but it must be remembered that every one is not so quicksighted, and that there are many to whom everything self-evident is not self-evident; whom also nevertheless it is desirable to assist in any efforts they may be willing to make in order to understand this subject ; and with this purpose I proceed to a close analysis of the LAW in question.

I have already drawn attention to the obvious facts of Nature here hitherto either controverted or neglected— (1) that every *point* of two different areas, equally exposed to the same source, has the same amount of Light upon it ;

(2) that what is true respecting the spreading and thinning out of Heat or Air is not true of Light; (3) that Light, from its nature, cannot spread at all; and (4), that even if its nature admitted of its spreading in this way, and deviating from the straight line between the source and object, there would be, in the present case, no vacant space for it to spread in,—no space unprovided with its full complement of Light for any more of this same Light, from this same source, to deviate into or to spread into, and so, to become diluted; also (5), I have shown that the Theory of the Inverse Squares, instead of being, as is pretended, the Law of Areas in Geometry applied to Optics, is precisely *the contrary* of this Geometrical Law.   I have now further to point out, what, as I say, many will not require to read at all, and most others will discern at a glance, that this Law of the Inverse Squares,—a proposition utterly different from the Theory of the same name, but equally devoid of all geometrical principle,—is something entirely without foundation of any kind, as entirely so as the Theory itself, and, like it, completely destitute of Data, Logic, and Common Sense; this, also, so obviously so, that it is inexplicable how it came to pass that so fanciful a "Law of Nature" could ever, among an enlightened people, have been received as this has been.

Two Interpretations possible.—The *soi-disant* "Law of Nature" however, here in question,—"the Law of the Inverse Squares" as it is confusedly called,—is to the effect that irrespective of all Medium, the degree, intensity, or quantity of Light, as well as that also, it would seem, of gravitation, *diminishes as the square of the distance from the source increases;* not, however, on account of this distance from the source, but on account of the greater space and surface supposed to be presented to the source at this greater distance from it.

Now, the original expression of the Law here given, viz., "that Light diminishes as the square of the distance from the source increases," can mean either of two very different things. It can mean either that the Light itself becomes less, or that the Diminution becomes greater in this ratio of the distance from the source (an effect, we are told, of the larger area which, at the greater distance, is supposed to present itself to the source); and so little attention has hitherto been given to this subject that I have found scientific men of European celebrity imagine, on the subject being first proposed to them, that the two principles now stated (the Diminution of Light, and the increase of Diminution) mean the same thing, and that when the Diminution is four times greater, the Light is four times less. It is easy, however, to see that this confusion, as well as, no doubt, the Law itself, must have been, in the case of such men, a mere momentary oversight.

Let it therefore be distinctly understood that it is only in the first of the two senses above indicated,—of the transmitted, not of the untransmitted Light,—that the Law is ever interpreted by scientific men. In the first only of these two senses therefore is it necessary to point out this utter absence of all Logic and of all data. It will be found, however, that by keeping at the same time the import of the second interpretation steadily before the mind, and comparing the two as we proceed, we shall be greatly assisted in understanding the import of the first.

To continue, then : The common interpretation of the Law among the Professors of Physics and Astronomy is not by any means, as it might so easily be supposed to be, that Neptune's orbit, being 30 times farther from the sun than ours, the whole diminution (or loss) which takes place, in the solar light, between the sun and Neptune, is (30×30) 900 times greater than that which takes place between the sun and the atmosphere of the Earth. As far as the mere words are concerned in which the Law is usually expressed, the Law *might bear* that interpretation; but this is not the one ever given to it; nor even the subject here understood by professional men, as intended to be treated of under the name of their Law " for the diminution of Light by distance.'

The received Theory upon this point, and the received interpretation of the Law as " the Law of the Inverse Squares," is that the solar light which falls upon each square foot of Neptune's atmosphere is 900 times less,—less strong, less intense, less in quantity, efficacy, and degree,—than that which falls upon each square foot of the Earth's atmosphere; —is, in short, as I have said, only one 900th part of it ; and this, it seems, although Neptune is not 900 times a greater area than we are (as the geometry of the cone would require him to be), but because he is 30 times farther *from the Source.* This interpretation of the Law, as a Law of the Light transmitted, is the one which is always given to it by those scientific men who write or have written on the subject. It supposes the luminous force of the sun, which we have in action upon each point at the Earth's orbit, to be divided into 900 equal parts or degrees, and maintains that one of these parts or degrees,—one of these nine-hundredths,— is the strangely limited amount of this solar force which reaches each point at the orbit of Neptune.   These writers think that this is the original import intended and the true application of that Law according to which Light is said to diminish in the same ratio as that in which the squares of the distance from its source increase. (*See* Appendix, Nos. 2, 3, and 12.)   They think that after the first distance there is thus 4 times more light than after the second distance, 9 times more than after the third, 16 times more than after the fourth, 900 times more than after the thirtieth distance ; and that, when we come to greater multiples of the distance, the light after the first distance is a million times greater than that after the thousandth distance, and a billion times greater than that after the millionth distance, *however short this first of the equal distances may be.* So that, whether the whole distance, so divided, be a mile, or a million miles, or only a single yard, and the divisions made be into miles or inches, the light at the end of the second division is thus *always* one quarter of that at the end of the first division, and the

K

light at the end of a yard, a millionth or even a billionth of that at the beginning; and this interpretation of the words they have now, happily for truth, secured to their Law by the introduction of the word "inverse." These writers and lecturers think that, since Neptune is 30 times farther, *not from the apex of our cone* (*i.e.*, of the cone of which Neptune's disc and ours are sections), but simply 30 times farther from the sun than we are, our discs must also accord in the geometrical proportion belonging to that *difference* of distance (which, however, it is known they do not; Neptune's disc is not 900 times greater than ours); and not that only, but also that, in consequence of this *supposed* geometrical proportion between our areas and our distances, the Light which falls upon each point of *all* space at Neptune's distance must consequently be 900 times less than that which there is on each point of *all* space at our distance; nay, further, they think that, on that account, although the source of Light is the same for both planets, and the whole amount (in this absence of all medium) admitted to be the same on both, yet the luminous energy in action upon each point of Neptune's atmosphere must, on account of his relative distance from the sun, and *therefore* (so they reason) on account of his relative area, or section of the cone, produce 900 time less effect, by this geometrical Law of sections or areas, than the luminous energy in action upon each point of ours; and accordingly, they limit the Law to this one of its two possible interpretations (to the interpretation of Light transmitted rather than of lost light), by introducing the word "inverse" or "inversely" into their expression of it; saying that the transmitted Light increases or decreases in inverse ratio to the square of the distance; which means that the one thing spoken of is less in proportion as the other is greater, thereby precluding all possibility of the law being interpreted of the diminution, or Light lost, this diminution being always a thing proceeding in a *direct* ratio, *i.e.*, more or less as the distance is greater or smaller.

For most people it is not improbable that the extraordinary differences of Illumination just alluded to, as resulting everywhere throughout the system from this Law, and never hitherto taken the least account of by the Profession, will alone be sufficient to make it clear that there must be an error somewhere, either in the Law or in its interpretation. That is therefore a useful preliminary. I now propose to analyse this Law, and to show that it does not treat of and could not treat of,—could not, one feels disposed to say, have ever been intended to treat of,— proportions in the Light transmitted, as in the instances just mentioned, but only probably,—although also most erroneously and unaccountably,—of proportions in the Light lost or untransmitted upon these occasions (*i.e.*, of the dark cones in the Spoke-Theory, PART II., Section 4),—an equivocation and absence of distinction in the terms of the Law, wholly overlooked, and never even once alluded to by any of the writers in question. This utter impossibility of the ordinary interpretation is what I propose now in this PART III. to deal with, irrespective of all other considerations, for it is independent of all others, and greatly assists all others. If it is once seen that the Law cannot, by any possibility of Logic, bear the interpretation always given to it by the professors of physical science, as "the Law of the Inverse Squares" any better than it could bear the other interpretation, viz., that of the Direct Squares and the untransmitted light, we may hope that, with the help of a little time, even the more dogmatical of us will have, in this circumstance, a further reason for suspecting strongly that, to use a moderate expression, this whole department of Physical Optics requires revision. I must, however, be permitted to say that this gentle judgment is far from being the full import of what has, upon this subject, presented itself as the fact to thinking men. The Law involves so manifestly an *à priori* impossibility,— there being no *data* whatever nor even semblance of *data* for its construction,—that one only wonders, as I have said,

K 2

how it ever came to be entertained an instant among the educated ; nor is it easy to look upon it otherwise than as originally the clandestine introduction, by a few theorists, of a notion which all writers now admit to be but a mere " opinion" (see Appendix, Nos. 3, 5, 7, 14, 18, 19), unfortunately favoured and encouraged from the first by being always expressed in the vaguest and obscurest and most deceptive language which it was possible to devise, and which, as above explained, has only been rendered definitely and distinctly false by the occasional and now more frequent insertion of the term " inversely."

But the importance of this distinction between the Light lost and the Light transmitted,—a distinction which the theorists do all they can to obliterate,—must not divert the reader's attention from the other and equally important distinction respecting the equivocal import here of the word " distance," which I have already so often mentioned.

All who understand the doctrine of the Inverse Squares know that, as explained in PART I., it is the enlargement of the object's area which, in this doctrine, causes the diminution of Light as well as the ratio of this diminution ; and all who know this know that it is not the distance from the window or other source, that makes one object or area larger than another.

To conciliate, however, the strange misapprehension of so many upon both these points of the doctrine, and to simplify for them the statements to be now made, we shall here suppose, as they do, that there is the Sun or some other source of Light at the apex of the pyramid alluded to in the Law, as determining the relative sizes of the objects illuminated, thus, for the convenience of their understandings, making distance from the apex identical with distance from the lamp, window, sun, or other source, just as they suppose it to mean, although as said above, we all know that the relative sizes of objects are not determined, and do not depend upon, their relative distance from the window,— or their relative distance from any other source of Light.

Use of the term "Inverse."—Here then we are speaking of
a luminous point, and of three or more equal distances in a
straight line from that point; and we are told that this
Light diminishes as the square of this distance increases,
without any reference to what it is which enables distance
to have this effect. As far as the mere words of the Law,
thus unaccompanied by the term "inversely," are concerned, they, as has just been mentioned, may here signify
either of two *wholly* different things; viz., they may mean
EITHER that the degree of Light at the end of the first distance is 4 times greater than that at the end of the
second distance, however near those points of distance
may be to one another; and 9 times greater than that at
the end of the third distance, also however short these
equal distances may be; OR the words of the Law may
mean that the amount of diminution produced by the first
distance, alone, is 4 times less than that which is produced
by the two first distances together, and 9 times less than
the diminution produced by the joint action of three distances. If we only regard the language used, we discern
at once that this language can bear either of these two
interpretations, in the absence of the term "inverse;" but
that when this term is used, it is only the first of these
two interpretations that is intended; because it is only the
transmitted light, not the untransmitted, that is greater
when the distance is less, and less when the distance is
greater. The least reflection, however, shows us that the
statement given in this first of the two interpretations
never could, under any combination of circumstances, be
true in fact, nor is it easy to imagine that it was ever even
contemplated by the original authors of the formula, however ready they may possibly have been, speaking only, as
they did, of what happens in a medium, to regard the Law
as true in its second interpretation, when the medium is
uniform; viz., in that interpretation in which it can be
understood of the "direct squares" and of the untransmitted light, or "Diminution;" which latter would thus be

made to increase directly as the square of the distance itself increases;—so that twice the distance would give 4 times the diminution and 30 times the distance 900 times the diminution ;—a sense of the words, in the law, which we easily confound with the other sense, in which other we speak of the Light itself, not of its diminution; and say that the Light *decreases* as this square of the distance increases, *i.e.*, becomes at once reduced to $\frac{1}{4}$, $\frac{1}{16}$, $\frac{1}{900}$, &c. of the quantity left remaining after the first reduction, when the distance from the source is twice or 4 times, or 30 times this first amount of distance which we choose to take, and from which the first reduction results.

SECOND SECTION.

LOGICAL IMPOSSIBILITY HERE OF THE TERM " INVERSE ; " NO DATA FOR IT.

THE special point here to be explained is that the Law, however impossible in that sense also, could only have been intended originally as a law of diminutions,—not of things diminished; and that even if the geometrical law of areas, in the contrary or inverted sense of it, were, as is alleged, true of their illumination, instead of being so clearly and experimentally false, it is impossible that the fact should ever have been known to us. We have no date nor accessible source of information on the subject nor the slightest grounds for supposing that the current opinion is correct; for even its advocates admit that it is a mere opinion which is expressed in this law (Appendix, Nos. 3, 5, and 7).

What we do not know upon this Subject.—The quantity itself, out of which the diminutions or reductions are taken, is, to this law a quantity wholly unknown. There is none ever specified,—none that can be specified. Nor have we any notion even of its proportion to the diminution which in any case it undergoes; nor do we know what, in any case, that diminution amounts to. Here then is an amount of ignorance, on our part, that virtually terminates the whole question.

No such law, as this pretends to be, could ever have been put forward by the first writers, nor ever, without complete inattention or the requirements of some absurd Theory, have been advocated by the more intelligent of our modern professors.   No law could, at any stage of the decrease, in this case, have given even the relative measure of the quantities remaining,—the sole quantities that are here in question. The Original Amount of the light, the Reduction made in it, and the Remainder are, in every instance, hopelessly hidden from us by the Medium.   This is manifest in the case of the Planets and the sun. The only amount of the solar light that we know of, is that which remains, and which we experience, after the two reductions made in it, *first* by the medium existing between the sun and our orbit or Atmosphere, and *secondly*, by our Atmosphere itself. We do not know what proportion either of these reductions bears to our amount of the solar light on the Earth's surface, nor what proportion the original undiminished solar light itself bears to the amount of it we experience. The only knowledge to which any one has ever pretended respecting either of these reductions themselves, or the reduced quantities remaining after them, consists of the two following points of relative knowledge: (1) that equal lengths of Ether (the interplanetary medium) produce nearly equal amounts of reduction in the solar light; so that 30 times a greater length of this ether (*i.e.*, distance in it) than the first length, or distance, would produce nearly 30 times a greater reduction than the first reduction alone amounts to ; and (2) that the reduction produced in the solar light by the atmosphere of one planet is, probably, nearly the same as that produced by the atmosphere of another. Beyond these two facts, such as they are (and they are very little), we know nothing respecting even these reductions.   Above all, we know nothing whatever about any other supposed sources of reduction.   What is our ground for supposing any other ? We have none. The professors tell us, indeed, that they have "made up their minds" to consider that there is one other source. (Appendix, Nos. 3,

5, 7.)   Why, then, do they not state their grounds for this "opinion"?   So far are we from having any grounds for supposing any other source of reduction except the medium, that we do not know whether, notwithstanding the reductions now mentioned by the interplanetary medium, the solar light may not be regarded as increasing instead of decreasing, as the ethereal distance increases, and precisely in consequence of that very extension of the solar rays which thence results, and which has been even thought to involve its diminution (*see* p. 110).

.. The foregoing considerations respecting our ignorance of these matters may be thus stated in other words: We do not know how much of the sun's light our atmosphere absorbs; nor do we know what amount was withdrawn by the Ether before it reached our atmosphere.   We therefore do not know, even by inference, the original or undiminished amount of the sun's light.   We know nothing here whatever but the amount of this light which remains for us after these two reductions of unknown amount, viz., that by the absorption of our Atmosphere, and that by the absorption of the interplanetary medium; the former being, we have reason to think, much the same for each of the planets, and the latter almost in proportion to the length of Ether which the light has traversed.   Our question here, however, it is well to remind the general reader, does not relate at all to the solar light upon the surface of the planets themselves, but only to that amount of it which falls upon the surface of their various atmospheres, or, as it may otherwise be expressed, the solar light as it exists at the different orbits, prior to the diminution of it effected by each atmosphere; and with regard to this amount of illumination beyond our atmosphere, we, as already observed, know nothing.   Not knowing either the original amount of the sun's light, nor the reduction effected in the passage, we cannot know the amount of this remnant which reaches either our planet's orbit or the orbit of any other planet.   All the reductions (*i.e.*, quantities withdrawn or untransmitted) are wholly unknown except in their relative proportion *to one another;*—*i.e.*, wholly unknown in their relative proportion either to our known amount of light on the Earth's surface, or even to the original amount of the solar light. The remnants, therefore, which reach the successive orbits are as little known as the original amount itself of the solar light. We cannot even know their relative proportion to one another. We can only judge of their relative reductions. In a word, we neither have the original quantity in question, nor the quantity withdrawn on each successive interplanetary diminution, nor the remainders. How then, I repeat, would it be possible for us to know that the light remaining after the second reduction is a quarter of that remaining after the first reduction? Can anything exceed that in unreasonableness? Or how can we know that the light remaining after this first reduction is reduced to one-nine-hundredth part of its amount, after the thirtieth reduction has taken place? Where, with our scanty knowledge, is the Logic or Common Sense in such reasoning, or in such a Law as this?

What we should require to know in order to frame this Law. —What is here essential to observe is that this Law of the Inverse Squares speaks only of the Remnants of Light which survive each diminution, and only of their proportion *to one another.* It does not speak either of the original amount of the undiminished Light, which it does not, in any case, profess to know anything about, nor does it speak of the diminutions which take place in it. Nevertheless in order to be able to speak of the proportion the Remnants bear to one another, we should, as observed above, know, *first*, what proportion the original Light bears to some amount we know, and secondly what proportion each reduction bears to this known amount. We should then be able to say what multiple the first Remnant is of the second or of the thirtieth; but not otherwise; and, as just explained, we have no such knowledge; no one pretends we have. The Law acknowledges that it knows as little about the first Remnant, in the case of Lamp or Sun, as it does about

the original amount. The media alone prevent this knowledge even if there were nothing else to do so. Yet the
Law proceeds to teach that this first Remnant is four times
greater than the second. But upon what grounds does it
make this assertion ? It clearly has no *data* for it,—no *data*
here authorizing the statement that the second Remnant is
a quarter of the first. Our ignorance here is manifestly too
complete to admit of such a Law as that thus sought to be
promulgated. In order to know this proportion between
the Remnants of a diminishing Light, we should, I repeat,
require to know the exact degree which exists previously
to all diminution, and in each successive instance the exact
degree or amount of this lost Light itself; neither of which
conditions exists in the case of the planets; and even in our
atmosphere they rarely exist, in any case, prior to experiment. We can therefore rarely say, in any case, even in
our atmosphere, prior to experiment or, at least, hypothesis,
that the degree of Light transmitted to one point is a
certain number of times greater or less than that transmitted
to another. In the case of the Planets, we never can say
this, either of the Light as under the action of the Ethereal
medium, or of the Light as supposed independent of this action.

What we do know upon this Subject.—All that we here know
is, as already said, strictly limited to diminutions, and to these
only as results of Medium. We know of no other diminutions
of Light in its passage from one point of space to another. We
only know Light as it exists in a Medium, and as it exists after
the action of a Medium. All assertion respecting it or its
diminutions independent of a Medium is entirely gratuitous
and fictitious. We have no such knowledge. All writers both
ancient and modern—some more—some less distinctly—acknowledge this. The first writers who spoke of the square
of the distance in connection with this subject, did so only
with reference to the diminution which happens in a Medium,
saying that Light in its passage through the air diminishes
as the square of the distance increases (Appendix, Nos. 11
and 12); and the moderns also always illustrate this Law

of the Squares by what happens in a medium. (Appendix,
Nos. 2 and 3). Now, what we know respecting the diminution of Light in a Medium is that the same cause produces
the same effect, and that in a uniform Medium, such as all
these writers always mean, equal distances give equal
reductions. This obvious fact of Nature, it will be seen in
PART IV., as already mentioned (p. 15), the professors deny.
But they admit sufficient for our purpose here. They admit
that in a very rare medium, like that which there is beyond
our atmosphere, equal distances produce, as a fact, very
nearly equal reductions, so that we can speak of the reductions resulting there from several distances, as being this
number of times greater than that resulting from only one;
and we know that if there are 30 such reductions, their sum
is 30 times greater than the first of them, or if there are
900 of these equal reductions, their sum is then 900 times
greater than any one of these equal reductions, the original
amount of luminous power being at last, 30 times or 900
times *more reduced* than after its first reduction; not 30
times nor 900 times less, as the theorists suppose it to be.
This is what we know upon this subject, and all we know.
The earlier writers thought, however, either that they
observed a more rapid diminution to take place in Light at
the greater distance from the source than at the nearer
distance from it, or that it was most probable the *divergence*
of the rays was attended with some such effect; and so,
we find it assumed in published writings, first by Bouguer
in 1729, that this diminution,—the cone of Darkness,—
increases when the distance from the source increases, and
of course in the same ratio as the sections of the cone or
pyramid; *i.e.*, not merely in the ratio of the distance from
the source, but in the ratio of its square. (Appendix, Nos.
11, 12, and 13. *See* also pp. 100–102.) We likewise find
that by Lambert, in 1760, it was taught with still greater
clearness, as a fact connected with our atmosphere (" the
passage of Light through the Air," as Lambert calls it),
that the square of the distance from the Source, is the

ratio in which this second diminution in every Medium
goes on (for these writers frankly stated that there were
two such diminutions) the diminution thus becoming greater
and greater for each length of Medium, as these lengths
were measured off more and more distant from the source;
so that after the second equal distance, or length of
medium, the diminution is (2×2) 4 times greater than
after the first, and, after the thirtieth distance, 900 times
greater than after the first; a comparison this of the supposed successive reductions in successive lengths of medium,
which, in order to make way for the "opinion" respecting
the Inverse Squares (on which a few theorists had made up
their minds) seems to have been inadvertently and confusedly transferred from the mere reductions made in a
quantity, to the quantity itself in which the reductions are
made, and from what happens in a medium to what is supposed to happen where there is none. But even if the
earlier conjecture were correct, we could not deduce the
second from it. Even if we could discover these 900 equal
reductions in 30 equal lengths of Medium, we could not on
that account say nor know—we could not, merely on that
account, mean to say—that the light transmitted to the
orbit of Neptune is diminished to one 900th part of that
transmitted to ours. We could here only mean to say that
the solar light at Neptune's orbit is 900 times *more reduced*
(not 900 times less) than ours is; that the degree of light-
giving power lost between the Sun and the Atmosphere of
Neptune is (30×30) 900 times greater than that lost, or untransmitted, between the Sun and the Atmosphere of the
Earth; or, to vary the expression of it, that the quantity
of the solar light or the degree or intensity of the solar
light—call this as we may—which exists at Neptune's orbit
is the same as that which exists, at ours, *minus* 900 times
the unknown amount of diminution which has taken place
between our Orbit and the Sun.

It will be seen from the foregoing pages, that there are
here two things to be most carefully distinguished, viz., the

light lost and the light transmitted,—the Reductions and
the Remainders. What it is important to attend to is that
*the equal distances give us the relative proportions of the one
light but not those of the other,*—of the lost light,—not of the
light transmitted ;—that it is therefore only to the former of
these two,—to the untransmitted or lost light,—that, however erroneously, the Law in question could ever possibly
have been supposed to apply by those first writers who
were attending to the subject; and that it is merely
through an error of inadvertence, or through the prejudices of Theory, that professional men of influence have
so long applied it to the second, viz., to the light transmitted. It is quite clear that as soon as such men can be
prevailed upon to attend to the precise point here in question, they will be utterly'astonished at the huge blunder
they have so long unguardedly made, and the professional
οἱ πολλοί will then of course follow them.

To recapitulate then this analysis of the Law called the
" Law of the Inverse Squares," constructed to aid the
Theory which underlies it, and which has been already
fully explained in PART II., it must be remembered that we
only know the degree or quantity of the Sun's light which
remains to us *after* the reduction effected in it by our
atmosphere. We neither know the degree which reaches
our orbit, nor the degree lost in the passage to our orbit
from the sun, nor that lost between the sun and any other
orbit; nor do we know the degree of Light belonging to
the solar action prior to all diminution. I ought perhaps
here to mention that what we mean in this place when we
speak of a degree of light as known, or not known, is that
we know, or do not know, how many times our own light,
or what fraction of our own light, the light we speak of
amounts to. We know then none of the things mentioned.
We only know that, the degree of light left for our use
after the enormous but unknown reduction effected by our
atmosphere, in addition to that unknown reduction effected
(before our atmosphere is reached), by that infinitesimally

rarer medium, the Ether; and we know that the orbits of
the superior planets have less of the Sun's original light
than that quantity, to us unknown, which our orbit has.
This is all we know. How then, with this limited knowledge, could we possibly pretend to say what proportion
the unknown amount of the whole original solar light,
minus the unknown portion not transmitted to the end of
the first distance taken, bears to this Remnant *minus* some
still greater but unknown portion not transmitted to
another distance twice as remote or thirty times as remote? How can we say anything, for instance, so definite
as that one is four times greater or 900 times greater than
the other?—that one is a fourth part or a 900th part of
the other? How can we say what proportion the unknown quantity $x-1$ bears to the unknown quantity $x-2$?
—and *a fortiori* what proportion the unknown quantity
$x-y$ bears to the unknown quantity $x-z$?

It seems impossible that even to a person unversed in
science, there can here present itself any difficulty. We
know that between the sun and our orbit there is some
decrease in the original amount of the solar illumination;
and it is possible to imagine,—although not to understand as
a fact,—that the unknown amount of that decrease between
the sun and our orbit might increase, for the superior
planets, in that special ratio, which we call the square of the
distance, so that the unknown amount of this Reduction at
Neptune's orbit might be 900 times greater than the unknown
amount of it at ours. I do not say it is possible to see
that it so increases, nor even to comprehend *how* it could.
I am far from saying this, although I have found some
professors look upon it as a very natural blunder that any-
body might make, and that I was making. I only say it is
possible to see that it might be *conceived* to do so,—that
for many minds there might be a *primâ facie* appearance of
it. This ratio,—this proportion of increase in the *diminution*, or untransmitted light,—is clearly *what* the Law was
originally intended to teach, and *all* it was intended to

teach, because all it pretended to have the *data* for teaching. It is quite clear that there is not the least foundation for the subsequent introduction here of the term "inverse," and the transfer of these proportions from one quantity to another, as effected by this term,—from the light lost to the light transmitted,—from the successive diminutions which take place between the sun and the planets to the successive amounts of light resulting from these diminutions;—a transfer made under the strange impression that this was applying the geometrical law of areas to the diminution of Light by distance (Appendix, No. 8), not the least foundation, I repeat, for saying, as so many of the most distinguished scientific men do here say, or seem to say, and have long taught or seemed to teach in their lectures and their books,—that when a quantity undergoes 900 equal diminutions in succession, the whole of these diminutions together being then 900 times greater than the first of them, the quantity remaining after all the diminution is therefore 900 times less than the quantity remaining after the first (for in no way do they suppose these quantities known but by these reductions);—not the least foundation for saying that, since the diminution of the solar light at Neptune's orbit is (however minute or vast this diminution may be) 900 times greater than it is at ours, it therefore follows that the solar light itself, otherwise wholly unknown in amount, which is transmitted to Neptune,—the luminous force of the sun at Neptune,—is the 900th part of what it is with us.

THIRD SECTION.

ILLUSTRATIONS OF THE ABSURDITY LATENT IN THE LAW OF THE INVERSE SQUARES.

I HAVE thus far endeavoured to explain that the Law of the Inverse Squares for the diminution of Light, so blindly and generally adopted by the scientific, is the mere fiction of a few Theorists, without the slightest pretension to any foundation either in Physics, Geometry, or Logic ; and the truth of what I have said may also be rendered perfectly manifest to every intelligent person by any one of the four following very simple facts, among many others that could be mentioned :—

1. One of these facts is the obvious one already alluded to, that if we understand the proportions spoken of in the Law to be, as the term " inverse " denotes, proportions in the light transmitted, instead of in the untransmitted or lost light, the immediate result of the Law so interpreted is that the light at the end of the first equal distance is *always* four times greater than the light at the end of the second distance, whether that second distance be an inch or a mile, or a million miles! the degree of illumination at the further end of any given distance always one quarter of that at the beginning of it!

2. Another fact resulting from the Law of the Inverse Squares, and to which also allusion has already been made, is, that by dividing any given extent of the solar system into parts sufficiently minute, we shall be able to have the solar illumination of any parts we choose, not only 900 times, but millions (nay, billions, trillions, nonillions) of times greater than the illumination of other parts; for although, by a division of Neptune's distance into 30 parts, we find him, by this law, to have only 900 times less of this light than in various other parts of the system round the sun; yet, if we divide that same distance into 1,000 parts instead of 30, we shall have the light at Neptune a million of times less than it is at other points all round; and if we divide Neptune's distance from the centre of the system into a million of equal parts, we shall have Neptune's share of the solar light a billion of times less than the solar light elsewhere, and so on *ad infinitum*.

Nor is this all; for although with this division of Neptune's distance from the centre of the system into a million of equal parts, we thus have the illumination, at a point near the centre, a billion of times greater than at Neptune's orbit, yet if we divide this same interval (Neptune's distance from the centre) into only two equal parts, instead of into a million such parts, we shall then have the light in the middle of the whole distance only four times greater than that at Neptune's orbit,—a very small difference of light in proportion, for so vast a distance as 1,500 million of miles ; and so on throughout the whole system.

: Nay, for those who desire it, the absurdity of this Law can be brought also into view by subdividing this last half of Neptune's whole distance from the sun (which last half derives all the light that travels through it from the amount of light at the commencement of it) into 1,000 equal parts ; when lo ! according to the Law of the Inverse Squares, we have Neptune's share a million of times less than that near the half-way point of his whole distance from the sun, instead of being a fourth of it, as was the case when the whole distance was divided into only two parts !

3. Another of the four facts which I here propose for the consideration of the scientific is the following : At a distance from a lamp or other luminous body, take any point in which the light is found by the photometer to be diminished to one quarter of the original light. Then bisect the distance extending from that point to the lamp ; and according to the ordinary interpretation of the law in question, the degree of light at the end of the whole distance is a quarter of that at the half distance, or point of bisection ; but by the hypothesis (or construction) it is also a quarter of the original light.

: 4. The last of the four facts to which I would here call attention as being any one of them instant and easy proof of this absurdity, is this : If we mark upon a rod, 3 feet long, 1,000 equal divisions, and place a point of electric light at one end of the rod, we have no difficulty in seeing that the illumination at the other end of the rod is not, as the law asserts it is, a million of times less than the illumi-

nation near the source,—*i.e.*, at the extremity of the first of
these minute divisions marked upon the rod. Nothing can
be clearer than this is with regard to the light transmitted.
No one will pretend to say that the light at one end of the
rod is a million of times less than at the other. Yet it may
be useful to the reader to observe that the Law, if understood of the reductions or relative quantities of lost or untransmitted light, might conceivably, at least *primâ facie*, be
true ;—as true with regard to these minute spaces as with
regard to the vast intervals between the planets. The whole
diminution in the distance of three feet must be supposed
to be a thousand times greater,—and might, as far as ap-
pearances go, be a million times greater,—than that which
takes place in the first minute division of the rod. It is,
I repeat, easy to see that, although the light transmitted
to the end of the rod is not diminished even to one-half of
that close to the source, nor perceptibly diminished at all,
nevertheless the whole diminution at the end, however little
it may be, is certainly 1,000 times greater than that after the
first of the 1,000 distances, and might very well, as far as
a *primâ facie* judgment goes, be a million of times greater
than that first diminution. Be that as it may, however, it
is quite clear that the light itself is not here diminished to
a millionth part of the original, which is all that this fact is
cited here to prove.

These four demonstrations of the absurdity into which
those Lecturers and other scientific Expositors have fallen
who all interpret the law as a law of the transmitted light,
apply in all their details to heat and to gravitation as well
as to Light, and it is unnecessary to multiply them. There
is no one, it appears to me, who will not find it perfectly
clear, from what has been said, that the thousand divisions
of the rod are not employed in the Law to indicate proportions in the degrees of light transmitted to the different
points of the rod, but only, however erroneously, proportions
in the various amounts of reductions or lost light caused by
the intervals (*i.e.*, in the amounts of light disappearing,

between the commencement and the end of each minute
division marked upon the rod), and that it could only be
through inadvertence that the principle of decrease thus
originally intended in the law, was without much distinctness, it is true, transferred by subsequent writers from the
one series of proportions to the other, having been afterwards however asserted distinctly and intentionally in the
interests apparently of a new Theory, by the introduction
of the expression "inversely" or "in an inverse ratio,"
simply to signify "in a contrary sense," or "in the opposite
sense."

"900 *times more reduced*" *never means* "900 *times less.*"—In
the same way it becomes also evident that when scientific
men look upon it as so certain that the solar illumination
at Neptune is 900 times "less" than ours, this, if we
reflect upon it, can only mean, however utterly and manifestly erroneous even that statement is, that his light is 900
times MORE REDUCED than ours,—that the reduction or lost
light in his case is 900 times greater than in ours,—that the
degree of the luminous force there, is, as already stated,
equal to ours *minus* the degree of this force lost between the
sun and the orbit of that planet.   But this does not mean
that he has only the nine-hundredth part of our degree of
the solar illumination.

Far from it. The very simple and obvious fact that when
we subtract even a known quantity from a quantity unknown, this gives no clue whatever to the quantity left
remaining,—in other words, the fact that a diminution of
light 900 times greater than the first diminution is not the
same thing as a remainder of light 900 times less than the
first remainder, nor even suggests the means of knowing
what proportion of the light remains after each diminution
unless we know likewise the original amount of light,—can
also be demonstrated by supposing a gas-light illumination
consisting of a million burners in a large hall.   If we extinguish 900 of these burners one after another, the diminution of light which has thus taken place in the hall (*i.e.*, the

amount of light lost) is 900 times greater after the last
burner has been extinguished than the diminution was which
took place upon the extinction of the first. But how can we
pretend to say that the light left in the hall is now 900
times less than it was before,—900 times less after the last
extinction than after the first? It is true that the original
light is now 900 times more diminished, more reduced, than
upon the first extinction, but not 900 times less, not a nine-
hundredth part of what it was at first. There still remain
999,100 burners not extinguished. How can that be called
a nine-hundredth part of the light existing in the hall after
the first burner was extinguished? Can anything be more
evident than it here is that to be 900 times more reduced is
not the same thing as to be 900 times less,—not the same
thing as to be a nine-hundredth part.

Here also we see an illustration of what was observed
respecting the absence of all data and foundation for
the law of the Inverse Squares. We see that it would
be impossible for us, at any point in the progress of the
extinctions, to deduce even the relative number of unex-
tinguished burners from the number of those extinguished
if we did not know, as we do in this supposed case, the
precise number that were in the hall at first. In like manner
we see that the relative degree or proportion of light transmitted to a certain distance could not be inferred from the
degree of it lost or withdrawn in the transit, unless we
know the original degree of it, as well as the exact degree
that has been lost. In the case of the planets, however, we
know neither of these things. We neither know at all, not
even relatively, the original degree of the solar light before
any diminution has taken place; nor do we know, otherwise than relatively to one another only, the degrees lost
between the centre and each planet. We therefore have
no basis nor data for the knowledge thus pretended to.

FOURTH SECTION.

RECAPITULATON OF PARTS II. AND III.

Thus far we have looked for the physical possibility of the alleged fact of Nature (Part II.) and the logical possibility of the alleged law of Nature (Part III.) ; and we have seen that neither exists.

We have seen that the true meaning of the Optical Theory respecting the Inverse Squares is that Light is inversely as the areas illuminated (less when they are large and more when they are small), although so many understand it to mean that Light is inversely as the Square of the distance from the Source, without any reference whatever to the area illuminated ; and we have seen that although the surfaces or areas of objects are enlarged by their distance in the cone or pyramid, they are not enlarged by their distance from a window or other source of Light. We have seen moreover that even if distance from the window or other source could make the object or area larger (either than it previously was or than the other object is, which we compare with it), there would still be the same degree or amount of Light upon each part of it as there is upon the whole of it. Its Light would not be diminished *upon each spot*, merely by its enlargement. We see that there is no geometrical foundation, nor any other, for the assertion that it is ; yet its being so diminished, upon geometrical principles, by the spreading, dispersion or dilution of each ray, or by its deviation from the straight line into unilluminated parts, is what is here asserted by the theorists, contrary to our common knowledge, and to their own admissions respecting the umbra and penumbra of the eclipse as well as respecting all other shadows. We have seen how fully the more enlightened of them, nevertheless, recognize the fact that, if we take no account of medium, the sun's rays pass undiminished in number and

intensity to the remotest limits of the system, and the other
fact that there is no point of the space confronting the disc
and exposed to these rays, to which they do not all come,
in the cases in which we cannot see this fact as well as in
the cases in which we can. We see therefore that even upon
the showing of these writers themselves, the sun's light is
not diminished by distance from the source, nor the larger
and the smaller space unequally illuminated; all that these
theorists here contend for being that the light which is
equal on the whole of each is unequal on each equal portion
of each, (" on each parity of surface,") the obvious fallacy
of which conception has been fully exposed (pp. 73, 77). We
have seen moreover that there is not the slightest pretext
for saying that the solar light does not *converge* from the
disc to each point of the space opposite as well as diverge
from the disc to the whole of this space; nor the slightest
pretext for the Spoke-Theory,—the Theory which teaches
that there are unilluminated spaces between the solar rays
which, becoming gradually wider, receive into their darkness a large amount of space in all parts of the system, but
especially in the more distant portions of it; and that, even
if light, when left to itself, and when without a medium,
did equalize itself as heat does, expanding on all sides until
it found its level, it would here have no opportunity of doing
so, not only because it is *never* left to itself, the Light passing always directly to each point from the source (and so
directly that, without a medium, no deviation from the
direct path is possible), but also (this is a very important
consideration) because there are no spaces on either side of
the rays, exempt from the same intensity of light, for
these supposed self-bending and self-diluting rays to expand
into.

We have likewise seen the Logical Impossibility of the
Law by which the light itself is said to become less as the
squares of the relative distances *from the source* become
greater; so that, to go no further, the light itself, after the
second distance, is thus represented as always being one

quarter of the light after the first distance, whether the second distance be a foot or a million miles. We have seen that with regard to the Sun and Planets we have not the slightest foundation in any knowledge that we possess,—not the slightest imaginable *data* of any kind, for such an assertion, inasmuch as we neither know the original amount of light before the first reduction takes place, nor the amount of either the first or second reduction, nor the amount remaining after either reduction ; we cannot therefore say that the light after the first reduction is four times greater than that after the second, even if in nature it really were so. We have seen that this extraordinary misconception of our facts results from the still more extraordinary misconception and belief that we are on this occasion merely applying to Light the geometrical law of areas (Appendix, No. 8) ; whereas, far from doing this, we here apply to Light exactly the contrary of that law, or that law inverted ; the geometrical law being that the area increases as the square of the distance does, whereas for Light, we invert this law and say the Light decreases as the square of the distance increases, an inversion of the geometrical principle which we hardly seem conscious that we acknowledge with the term " inversely "—meaning thereby, *in the opposite sense of the geometrical law.* There is reason to think that when the law of the Inverse Squares was applied to Light by the earlier writers, it was only thought of with reference to the Light lost, not to the light transmitted ; for in whatever way Light was here supposed to be reduced by distance, the law in question could be supposed to apply only to these reductions—very erroneously, it is true, but still only to these—not to the remnants of Light left after them ; to the " direct squares," as they may be called, not at all to the " inverse " ones. It was the diminution only, not the light itself which could possibly have been supposed to march, in any sense, *pari passu* with the enlargement of the Area. We see also, however, that even this application of the law to the Light lost could only be true, if it be true, which 'has never yet been shown, that to double the cause in nature is ever, in any case, to quadruple the effect; since in this theory the first step is that twice the distance gives four times the diminution; apparently also, that two suns would give four times as much light as one; and we see that even this application of the Law in question to the reductions would not give Neptune's illumination as one 900th part of ours, because it only represents the *amount abstracted* in the passage to his orbit as 900 times greater than that abstracted in the passage to ours; in other words, that the solar light is 900 times *more reduced*, not 900 times *less*, in amount, force, and intensity at his orbit than it is at ours. Finally we see that none of the alleged reasons exist for supposing that there is any second diminution of light in nature as an effect of distance from the source, *i.e.*, any other such diminution of it except that which is occasioned by its passing through a medium; and that all writers are agreed that the diminution of light in a medium does not proceed by the law whose flagrant fallacy we have here seen, but by another and very different law, much more like the common one, and which shall now be examined in PART IV., whereby it will be seen, as this Treatise undertakes to show, that the solar light is so nearly equal outside the atmospheres of all the planets, that, if we were in a position to make the comparison, even the most powerful photometers that we possess would not enable us to detect the difference of illumination that there exists between the different portions of the system.

END OF PART III,

PART IV.

THE LAW OF DIMINUTION FOR LIGHT PASSING THROUGH A
MEDIUM, AND THE EFFECT OF THIS LAW UPON
THE ILLUMINATION OF THE PLANETS.

PART IV.

THE LAW OF DIMINUTION FOR LIGHT PASSING THROUGH A MEDIUM, AND THE EFFECT OF THIS LAW UPON THE ILLUMINATION OF THE PLANETS.

FIRST SECTION.

DIMINUTION IN A MEDIUM THE ONLY DIMINUTION OF LIGHT IN ITS PASSAGE FROM ONE POINT OF SPACE TO ANOTHER.

We have seen, in PART I. and in PART II., that there is no such Law of Nature, even logically possible, as that of the Inverse Squares, nor, in Nature, either of the two supposed grounds for it, viz., either the Deviation and Spreading of Light independently of the Medium, or the Spaces without Light between the rays into which it is supposed Light spreads; and this disposes completely of the gratuitous hypothesis of these Inverse Squares, so persistently advanced by professional men for explaining the diminution of Light by distance from the Source; as if the simple and natural fact of Absorption by the medium did not do this sufficiently, as well as altogether preclude, by the constancy of its action, the possibility of our knowing any other. We shall now give a little attention to such peculiarities in this diminution of Light by distance from its Source (*i.e.*, by length of Medium) as apply to the illumination of the solar system; and from these it will be clearly seen that the whole system is equally illuminated. This the Professors themselves will be the first to discern, as soon as they are once able to see that

the Old Theory of the Inverse Squares is no longer tenable,
—nay, has been concealing from them one of the grandest
facts of Nature.   They will then see, without any further
difficulty, that all those other professorial theories about
"yards of Light," its "concentric spheres," its "dilution,"
"expansion," "ductibility," &c., fall at once to the ground,
as being, like the theory of the Inverse Squares which they
are intended to support, no less false than useless; and that
the solar light is manifestly not only not so unequally distributed throughout the system as they have hitherto been
led to imagine, but is not unequally distributed at all.   Their
knowledge of the details will make all this easy to such
men as soon as they have once mastered the utter impossibility and utter nonsense of the alleged fact and alleged
law on which they have been so long accustomed to depend.   It is mainly for the sake of the reader less versed in
these details that I add this Fourth Part to the Treatise, in
order to point out what it is, outside our atmosphere, that
really causes any diminution of Light that takes place there;
and what the law and conditions of this diminution are;—a
discovery, it must be admitted, which comes unaccountably
late, when we consider the vast amount of attention that
Optical Research has recently obtained.   This knowledge
will help him materially to see the truth of that equal
distribution of Light throughout the solar system which it
is the object of the present pages to make known, and will
also make clearer to him, on many points, the frivolity and
absurdity of the old law and its old theories, from which it
was represented as deducible that the Light on Neptune's
surface was a 900th part of that which we have on ours.

SECOND SECTION.

THE PROFESSORIAL LAW: ITS TWO OBVIOUS ERRORS.

BUT, before we enter upon the facts of nature as they here
present themselves before us, it is proper to point out the

curious blunderings of so many professional men even here
where one might have supposed it impossible to blunder.

The common facts of nature in this place are, as mentioned
already in PART I. that all Light passes through some
medium, and that all media reduce its intensity by Absorption, but more or less rapidly, *i.e.*, at a shorter or longer
distance, according to the density of the medium; the
obvious law of nature for this diminution being that, in
nature, the same cause produces the same effect; that,
when the medium is uniform, twice the length of it (*i.e.*,
twice the distance in it) gives twice the reduction; a threefold distance, a threefold reduction; and so on; that the
length of medium which absorbs the light of 20 candles will
absorb that amount of Light whether the source consists of
20 candles, or 500, or 5,000, or 5,000,000 candles; and
that if, in any case, half the Light is absorbed in the first
distance, the other half is absorbed in the second of these
equal distances.

These two simple and obvious natural principles are, as
there mentioned, both of them denied by professional men.
They say that the amount successively absorbed is greater
or less, according to the amount that is left by the preceding
distance, not according to the amount of Medium; and that
the absorption is never complete, never finished, however
long and however thick, the Medium may be;—two very
remarkable statements, it must be admitted, to originate
with men whose time and thoughts are wholly given to
these subjects.

Their Theory is that the absorbent powers of a medium
depend upon the degree, quantity, or intensity of Light that
enters it; that if this quantity be great, the amount absorbed
is proportionably great, and, if this quantity be little, proportionably little of it, also, is that which is absorbed.
Their law is that, whatever reduction is made by the first
equal length of a uniform medium, less reduction is made by
the second of these equal lengths; still less, by the third;
and so on according as there is less Light from which the

reduction, in each case, is to be made. The question will
naturally be asked: How much less? Their answer is that
this depends on the density of the Medium or on the length
of it; on the amount of Light, therefore, absorbed or abstracted in the first distance traversed. If the medium is
dense or extensive, and therefore absorbs much in the first
of the equal distances, then the difference between the first
reduction and the second is great; whereas if the Medium
is very rare even if extensive, or short even if dense, and
therefore absorbing little in each length, then the difference between the second reduction and the first may be
very little, and can in fact be even found to be wholly
imperceptible.

This Law is thus expressed: The Medium being uniform
and the distances equal, the Medium of each succeeding
distance absorbs, not the same amount of the original
Light as was absorbed by the preceding distance, but only
the same proportion or fraction of Light as that taken from
such Light as enters the Medium of each preceding distance, whether this *entering* Light be the whole original
unreflected Light, as in the case of the first distance, or
only a part of it, as in the case of every subsequent
distance.

I have already given an illustration of this strange Law
in PART I. The following is another instance:—

In a uniform Medium take any distance from the source of
Light, and divide this distance (or length of Medium) into
four, or any other number of *equal* parts, lengths, or distances, through which the Light passes. Now, if the first
distance reduce the original Light by one quarter, then
each succeeding distance will absorb one quarter;—not
however one quarter of the original Light (as ordinary
people hold), but a quarter of the *fraction* that enters it,
and that is left, in each case, after the preceding distance.
The second distance, then, will absorb a quarter of the
Light left after the first distance; the third distance will
absorb a quarter of the Light that is left by the Medium of

the second distance; and the fourth distance will do the same for the fragment of Light that has been transmitted through the third distance; and so on, to any number of these equal lengths or distances; less and less Light entering each succeeding length of Medium, and less and less being absorbed by each, because each length or distance absorbs the same *fraction* or proportion, but not the same *amount* of the Light which enters it.

In like manner, if the first distance absorbs three quarters of the original Light, leaving only one quarter of it to enter the next distance, then this next distance will absorb three quarters of that quarter, transmitting to the third distance only a quarter of that quarter, or one-sixteenth of the original light; out of which sixteenth part, the fourth distance will also take its three quarters; and so on, to any extent of Medium; there being always, by the theory, some remainder; and each succeeding distance, although equal, absorbing less and less, *on account of the light becoming less and less which enters each.*

So likewise if the medium is so rare, or its equal lengths, the distances, so short, that the first distance absorbs only a thousandth part of the original Light, then by this strange theory, the second distance will absorb only a thousandth part of what remains; and the third distance will abstract but a thousandth part of what the second distance has left; &c., &c. In this latter instance, according to our common notions and experience, at the end of a thousand equal distances, the whole light would be extinct; since with us a thousandth part is absorbed by each length of Medium; but not so in the theory we are speaking of; for in it, there is never even so much as a thousandth part of the original light absorbed by any one distance, except the first, although the distances are equal and the Medium uniform. In this case, therefore, at the end of a thousand distances, there would still be a large remainder; which, although thus, according to the theory, continually decreasing, would nevertheless last through any number of equal distances or any density of medium, AD INFINITUM. This must not be forgotten,—AD INFINITUM.

As to this latter point, which is the second principle involved in this theory, viz., the alleged impossibility of light being wholly absorbed by the Medium through which it passes, however dense and extensive this Medium may be, this second principle naturally follows from the first: Since, at each distance, there is deducted only a portion (never the whole amount), of the light that enters that distance, there never comes, never can come, a distance, or length of Medium, which absorbs the whole light.

These are the two favourite hypotheses of all professional men and of many others, respecting the diminution of Light by the Medium,—a diminution which obtains among them in the presence of the uninitiated, the mysterious but dignified, because mathematical, name of a " Geometrical Progression."

Now, what are we to think of the statement that the amount of Light which enters a given medium, determines the absorbent power of the medium, and therefore the amount which is absorbed? What sort of impression is such a statement, when distinctly put forward, likely to make upon people not imbued with the prejudices of the profession? Is it likely to increase their confidence in the Priests of Science? or, is it supposed that unprofessional people are all without Common Sense? For what does such a statement mean? If we have the light of two candles to deal with, and employ a medium which shall absorb half their light (i.e., the light of a single candle), we seem to learn pretty distinctly, from this circumstance, the simple fact of nature that the medium thus employed absorbs exactly the light of a single candle, neither more nor less. Now, instead of two candles, let us employ but one; and according to the theory in question, the absorbing power of the same given length of Medium becomes reduced to such an extent that it now absorbs only half the illuminating force of one candle; or if, on the contrary, we

increase this original light twelvefold and employ a dozen
candles, instead of one, then, according to this eccentric
theory, the absorbing power of this same medium is also increased twelvefold, and it absorbs an amount of Light equal
to half the light of the twelve candles; *i.e.*, it absorbs the
light of six candles. The *same* medium absorbs half the
amount of any light that enters it ;—can be made to absorb
in short any amount of light that you require.

If (for the sake of round numbers) you employ an electric
light equal to the amount of unreflected illumination represented by 1,024 candles, and divide your uniform Medium
into ten equal lengths or distances, then the first length
of it absorbs half the whole light, and there remains only
the light of 512 candles to enter the second length. In
that second length the light is further diminished, by
this law, to one-half of that which enters it, viz., to the
light of 256 candles. In the third of these equal lengths
of the same Medium, the 256 candles are reduced to
128; in the fourth to 64 candles; in the fifth to 32
candles; in the sixth length to 16 candles; in the seventh
length to the light of 8 candles; in the eighth length the
reduction is to 4 candles; in the ninth to the light of
2 candles; and in the tenth length, we find that the same
amount of Medium as can absorb the light of 512 candles
when the light of 1,024 enters it, cannot absorb more than
the light of one candle when the light of only 2 candles
enters it from the preceding length. And thus, any given
amount of Medium can be made to absorb any amount of
Light that is desired. The usual rush will no doubt be
here made at once by the superficial portion of the Profession to some of the technical hiding-places in Optics, or to
the fourth,—perhaps fifth or even sixth,—dimension of
space, or to some other similar absurdity, in order to be
thereby enabled to say something that shall seem to justify
all this incoherence; but the more enlightened members of
the Profession will instantly acknowledge the utter unreasonableness of such a Law. And can anything, we may

M

well ask, exceed such a Theory and such a Law in nonsense? Is it possible to believe that in England, at the end of the nineteenth century, such "scientific" dogmas as these are taught, and listened to, and paid for?

As to the second of the curious theories in Absorption, now under consideration, to the effect that no amount of curtains, or of London fog, or of atmosphere, or of water, or of any medium whatever, can *wholly* absorb the solar rays or any other rays,—even those of a rushlight,—it would be mere waste of time to point out, in "interesting experiments," the utter inaneness of such a theory. Every reader will discover this for himself; and many even of those very writers and lecturers who have so long refused to listen to us upon these points, admit of their own accord, although apparently without being aware of what they are doing, that this scientific statement of theirs is utterly false. They all teach that, at the bottom of the Ocean, there is no light whatever,—none at least, except that singular invention of theirs called "invisible light;" and we, all, simple-minded people, know that the most pervious Medium, if it is extensive enough (*i.e.*, if the distance is sufficient) can absorb any amount of light that we employ, —even "the light that is invisible,"—as well as that, if the Medium is sufficiently thick, a length of it, infinitesimally minute, is sufficient to do this. (Appendix, Nos. 1, 16, 20.)

THIRD SECTION.

THE OBVIOUS AND ORDINARY OR UNPROFESSIONAL LAW FOR THIS DIMINUTION IN A MEDIUM.

Let us now contrast the ordinary law with the foregoing and then pass on to the facts of nature to which both laws are supposed to refer. In this latter department of the subject also, we shall see some remarkable instances of toleration, if not of oversight, on the part of those who write and lecture.

I propose here to speak only of the interplanetary Ether as medium, and of the Sun as luminary. It is unnecessary that I should detain the reader with any others. The Law of medium is the same for all media and for all sources of the unextended or central character here in question. This Law in its obvious form and as ordinarily received by unprofessional people, is, as I have stated above, that, in Nature, the same cause always produces the same effect ; that the same amount of medium produces the same amount of absorption ; that in a uniform medium, double the distance doubles the diminution ; that treble the distance trebles the diminution, &c., and that any medium, if dense and long enough, can absorb the whole Light. There is here no longer any question as to twice the length of medium producing either more or less than twice the diminution,—more, as when doubling the cause, (viz. the length) quadruples the effect, or less, as when the second equal length of medium, although equally dense, has not the same absorbent power as the first ; nor is there in this simple interpretation of Nature, any question as to Light when it traverses a Medium, traversing also at the same time, a space exempt from Medium, and being, in that second kind of space, always 4 times less at one end, than at the other, of any given length, whether this length be an inch or a mile. This latter principle of decrease could only have been listened to with reference to so obscure a decrease as that supposed to result from the enlargement of the space or from the enlargement of a ray, and seems never to have been *distinctly* asserted by anyone, with reference to the decrease resulting from a medium ; although it is sometimes very difficult for us to know whether the lecturers intend this principle to be applied to the medium or not. Nay, even in this matter of mere *reductions* (in the light), the professorial theory with regard to a uniform medium, does not represent the increase of reduction as more rapid than the distance. The professorial theory does not even consider this increase to proceed as rapidly as the distance (which we unprofessional people

think it does), but frankly informs us on the contrary, as we
have just seen, that, according to its principles, the reduction
does not keep pace with the distance from the source; that
double the distance does not so much as double the reduction; and what is still more extraordinary, that this
reduction is always becoming less and less, instead of more
and more, as the distance from the source increases!—So
that, when, under the supposition that the interplanetary
medium is uniform, unprofessional men say that, since
Neptune is 30 times farther from the centre than we are, he
has therefore the solar light 30 times more diminished by
this Ether of space than we have, (in other words, that the
diminution of the solar light on reaching him is 30 times
greater than on reaching us), professional men tell us that
this is a mistake; that the diminution in a *uniform* medium,
becomes less and less in every succeeding distance, and that,
upon this principle therefore, of a uniform Ether, the diminution at Neptune, is not nearly 30 times greater than it is for
us. Thus whatever diminution our common notions give
us of the solar light at Neptune, as resulting from the medium
it has traversed in order to reach him, these theorists assure
us that this diminution is not really so great;—that, as far
as mere medium is concerned, Neptune's light from the sun
is much more nearly equal to our own than even we suppose,
—that if we disallow their theory of the Inverse Squares,
then the sun's light outside Neptune's atmosphere is not
nearly 30 times as much diminished as it is outside ours.
They will see however that without their law we also have
this result; and, that, even if we had not, yet the difference
of solar illumination resulting from the medium between the
two planets is so utterly imperceptible, that this advantage of their fantastic theory can well be dispensed with.

FOURTH SECTION.

HIGHLY RAREFIED NATURE OF THE INTERPLANETARY MEDIUM,
OR SOLAR ATMOSPHERE IN THE SPACES BETWEEN THE
PLANETS.

IT is now well known to scientific men, that the diminution
of Light, resulting from Absorption, which takes place in
any given length of the rare medium between the planets
and the sun, is at least 250 millions of times less than it is
in the same length of the ordinary atmosphere that absorbs
Light, or that we breathe, upon the Earth; inasmuch as
that thin medium requires to be at least 250 millions of
times less dense, less obstructive, or, as it is termed, less
absorbent, less substantial, in order to admit of the rapid
movements of the planets;—movements so rapid that, even
in the upper regions of our atmosphere, they would meet
with the same resistance, says Sir Isaac Newton, as if they
were attempted in a universal medium of "molten gold."
So fully is this fact recognized by men of science that the
space between the planets has been frequently supposed by
them to be a vacuum; a notion which has no doubt contributed not a little to the other notion that the diminution
of Light in its passage through this space must be conducted
upon some principle independent of a medium. That it
should be a vacuum, however, is considered by the more
enlightened to be as impossible as any density that could
obstruct the planets. I need scarcely say, after the explanations of PART II., that if Light could traverse a vacuum,
and if this were what it did in its transit from the centre to
the planets, there would then be no diminution of Light at all
outside the atmospheres of the planets. By far, however,
the more reasonable and general conclusion is that the space
in question is not a vacuum, but that the medium which
fills it, is of a nature infinitesimally attenuated,—partaking
apparently, almost, as much of this infinitesimal nature, as

the vibrations productive of Sound, Light or Colour are now
so generally supposed to do. "It has been computed," says
one writer, "that a *cubic inch* of the air we breathe would be
so much rarefied at the height of 500 miles that it would fill a
sphere equal in diameter to the orbit of Saturn." "The air
in proceeding upwards," says another, "is rarefied in such a
manner that a sphere of that air, which is nearest to the
Earth, but of *one inch* diameter, if dilated to an equal rare-
faction with that of the air at the height of ten semi-diameters
of the Earth (40,000 miles) would fill up more space than
is contained in the whole heavens on this side the fixed
stars. And it likewise appears that the moon does not
move in a perfectly free and unresisting medium; although
the air at a height equal to her distance is so many *millions
of millions* of times thinner than at the Earth's surface, that
it cannot resist her motion so as to be sensible in many
ages." (*See* Ferguson's "Astronomy.") And no one else
has written anything very different. Thus, then, as far as
Medium is concerned, there is nothing, or next to nothing,
to diminish the original amount or degree of the solar light
in its long transit to the most distant of our planets. The
difference which exists between the effect produced by a
length of medium thirty times greater than another length
of the same medium, although something, no doubt, very
perceptible, and also very great, in our dense atmosphere,
would, we can easily see, be completely inappreciable in a
medium where this difference would be millions,—nay,
according to some writers, billions of times less,—less perceptible, therefore, than it is in the lower strata of our
atmosphere. Any little diminution in the vast amount of
the whole solar light, resulting from a few cubic yards or
cubic miles of our atmosphere, even if all concentrated upon
its path through space to Neptune, would, I repeat, be
utterly and manifestly what the human eye could not
discern.

FIFTH SECTION.

CONFUSION MADE OR TOLERATED BETWEEN THE DIMINUTION
FROM ABSORPTION AND THE OTHER DIMINUTION WHICH IS
SUPPOSED TO BE INDEPENDENT OF ABSORPTION.

NOR is this substitution of 30 for 900, in consequence of
our having merely to deal with the diminution of Light in
a medium, the only point with respect to medium, which is
here entitled to attention.

It will be found that with few exceptions, the writers
upon Light and Optics do not mention in their Treatises the
diminution by Medium and Absorption, or the law supposed
to apply to it, whereas they all mention the other diminution
and the other law; yet this diminution of Light by a Medium
is the only one known or thought of, except by those more
advanced in Physical Science. All the rest of the world
think only of the action of the Medium, when they think at
all of this diminution or the cause of it. Although the
more enlightened of the Professors are aware of this fact,
and sometimes laugh at it as a popular error, from which
they are themselves exempt, they never, or very rarely,
in their books or lectures, explain that the diminution they
speak of is *not at all* the diminution by Absorption and the
Medium, but simply and solely a diminution by Expansion
or Dilution and by the enlargement of the space illuminated,
or, as they prefer to express it, by the principle of the Inverse Squares, while the class to be called δι πολλοί of the Profession seem entirely unconscious of there being any such
difference to explain. But however much Professors may
vary in information, they almost all write and lecture as if
there were but the one kind of diminution in question, the
better informed of them well knowing that, for their readers
and auditors, that one is the diminution by medium only.

When, accordingly, they say that Light diminishes in the
rapid ratio known as "the square of the distance" they
know that it is understood by all except those among

themselves who know better, as a result of Absorption and
length of Medium (this, moreover, in our dense atmosphere),
and that this medium, with the Law of the Inverse Squares
applied to it, is supposed to be what makes the solar Light
on Neptune's surface 900 times less than that which we
have on ours.

As if to confirm their readers and their auditors in this
error,—I do not say that this has always been the intention, but as if it were,—all the Professors, as has been
pointed out (PART II. sec. 6), undertake to explain and to
prove the alleged diminution by dilution, deviation, or
expansion, under the name of the " Inverse Squares," by
means of the diminution which results from a Medium
(Appendix, Nos. 2, 3, and 15.)   Yet how could one of these
processes of diminution prove or explain anything of the
other ?   All thoughtful men are expecting and waiting for
an answer to this question.   Those are deceived who think
that no one notices the absurdity here so patent.

It being commonly believed that, as there is 30 times
as much distance between the Sun and the surface of
Neptune as between the Sun and the surface of the
earth, there is therefore 30 times as much Medium; and
it being commonly taught that as Light diminishes by the
square of this "distance" or length of Medium (for the unprofessional world, I repeat, are here taught no distinction),
our amount of Light is 900 times greater than that of
Neptune, it is now the general impression among all except
those well versed in Physics, that there being 30 times
more medium between Neptune's surface and the Sun than
between the Sun and ours, is the reason why Neptune's light
is held to be a 900th part of that which we experience here.
When, however, it is once known that the only diminution
connected with distance that we have to deal with, is that
which results from the Medium, and when it is known that
the Professors themselves admit that this diminution by
Absorption is not only not more rapid than the distance
(which the other diminution is supposed to be), but is not

even as rapid as the distance,—when it is once known that
30 times a greater amount of Medium not only does not
reduce the amount of *Light* at Neptune to a 900th part of
what it is with us, but does not even reduce it to a 30th
part of ours,—nay, does not even give 900 times a greater
amount of *reduction* in the one case than in the other ;—
when all this is thus at length known, those who read and
those who listen will be astonished utterly at the strange
blunder in which they, with their instructors, have been so
long imprisoned.   It may even be worth their while to
ascertain whether their instructors were indeed always in
prison with them, or whether their instructors were not
sometimes their jailors.

SIXTH SECTION.

TWO OTHER ERRORS TOLERATED AND ENCOURAGED.

THIS confusion or identification, however, of an amount of
Light 900 times smaller, with an amount of *reduction*
30 times greater, and of a 30 fold reduction with a reduction 900 fold, are not the only points that are here
entitled to attention and comparison.   There are two other
remarkable errors hitherto tolerated, where not taught,
upon this subject, apparently (whatever may have been the
real object) for the purpose of giving plausibility to the
enormous reductions involved in the theory of the Inverse
Squares,—two errors which require in this place to be taken
full account of,—especially one of them,—also the two
facts of Nature which they distort, and which seem hitherto
to have entirely escaped the notice, even of the professional
public, in this calculation.

One of these facts of Nature is that, although Neptune is
30 times further than we are from the centre, the solar rays
do not pass through thirty times as much medium on their
way to the surface of Neptune as they traverse in their

passage to our eyes on the earth's surface. In the case of each planet, the solar light only passes through one atmosphere in addition to the amount of mere Ether traversed before it reaches that atmosphere. To traverse 30 times as much medium as it traverses in reaching us, it would require to traverse 30 atmospheres as well as thirty times as much Ether; and this it does not do.

The other of the two facts to which I now advert, as omitted in all the calculations on this subject, is that the ethereal medium through which the solar rays always pass is never uniform;—that the highly attenuated medium between the planets becomes, through the attractive action of the Sun, more and more attenuated, according to the increase of its distance from the centre of the system, and that, upon the commonly received principle of science, it does so at the enormous rate involved in the square of the distance. Here, then, we see that there is not even 30 times as much of the Interplanetary Medium, nor nearly 30 times as much of it, for the solar light to traverse in its passage to Neptune's orbit as in its passage to ours;—so that, although 30 times a greater diminution from so attenuated a medium would be imperceptible, here we have to deal with very much less than 30 times a greater diminution; and this, be it well remembered, in a medium in which all diminution is 250 millions of times less perceptible than it would be in our atmosphere.

SEVENTH SECTION.

CONCLUSION AND RECAPITULATION OF THE PROPOSITIONS INSISTED UPON IN THE TREATISE.

In conclusion, then, we see that we have not, on each spot of the earth's surface, 900 times a stronger Light (a larger amount of Light) than there is upon each spot of Neptune's surface. We see, moreover, that even the amount of diminution which takes place before reaching Neptune's surface is not 900 times greater than that which takes place before reaching ours. Nay, we see that, although Neptune is 30 times farther from the Sun than we are, this amount of diminution is *not nearly* 30 times greater between his orbit and the Sun than it is between the Sun and our orbit; and that (which is the most important point in this Last Part of the Treatise) the very small difference of Diminution, *i.e.*, of Light lost,—not of Light transmitted,— here in question, is 250 millions of times less perceptible in the Interplanetary Medium than we find it to be in our dense atmosphere. We have no difficulty, therefore, in fully recognizing that this small difference (not nearly 30 fold) between the diminution which our light has undergone and that undergone by the light of Neptune, is what would be entirely unappreciable, not only to our unaided powers of sight, but even with the aid of the most powerful apparatus or photometer not only that we possess or know of, but even that we can imagine.

I conclude with the following brief Synopsis of my Propositions :—

(1.) All the planets receive the same (appreciable) degree of illumination from the Sun, notwithstanding the difference in his distance from them; the whole of the solar system being thus equally illuminated.

(2.) Light is not, in Nature, diminished by Distance in more than one way; which is by Absorption in the Medium.

(3.) In Nature, the Light falling upon a surface from a source to which every point of the surface is uniformly exposed, is neither diminished nor increased by the enlargement of the surface illuminated, and thus uniformly exposed to it.

(4.) Light, in Nature, neither seeks its own level, nor spreads at all. When we take no account of the Medium, it neither deviates from the straight line, nor can exist apart from its source, nor can be " diluted."

(5.) When we take no account of Medium, the whole

expanse of the surface *equally* exposed to the source, *i.e.*, exposed to a point or central source,—has the whole of the illumination upon each spot of it, and this at all distances from the Source.

(6.) Taking thus no account of Medium, the smaller surface, even when it is at the greater distance from the Sun or other central source, is, in every spot of it, as completely exposed to the whole luminous force of the disc, as the larger surface is, even when this larger is nearer to the Sun.

(7.) The whole degree of Light being on each square inch which is exposed to it, does not make the square foot have 144 times more than the whole degree of Light to which it is exposed.

(8.) There are, in Nature, no rays which diverge from one another, with dark cones or spaces between them, widening as the distance from the source increases; what are called " rays " being merely artificial or imaginary divisions of Light without, in nature, any real intervals whatever between them.

(9.) In Nature, no Light is diminished *merely* by the contraction of the central source, even when this contraction is *real;* clearly then not by the merely *apparent* contraction of it.

(10.) The Sun's light which reaches us has been diminished by our atmosphere, and by the ethereal or interplanetary substance only,—not by Perspective or the seemingly diminished disc.

(11.) Doubling the cause does not, in nature, ever quadruple the effect, nor does trebling the cause, ever, in nature, render the effect ninefold.

(12.) Light is only measured by degrees, never by yards, nor inches, nor cubic measure.   It is only the object illuminated that can be so measured, or the area from which the Light proceeds, or the materials or vibrations employed to produce it.   We have not, in nature, a block of Light, nor a yard of Light, although we have both of these in books and lectures.

(13.) Each square foot, or inch, or mile of the solar disc does not give out a Light of its own which reaches the confines of the system, or even the planets, and which is thus independent of the rest of the disc.

(14.) The amount of luminous force in any given line of Light, or path, or direction, or ray, is not increased by the length of the path or ray ;—*i.e.*, by its distance from the source. This amount, or degree, in the absence of all medium, would be, it is admitted by all parties to this controversy, exactly the same in the whole path, taken together, and in each point of it.

(15.) We have no grounds whatever for the sub-divisions of the Solar Light or of any other light, which are supposed to result from the sub-divisions of the surface upon which it falls, as indicated in the Theory and Law of the Inverse Squares. (*See* paragraph 12.)

(16.) We do not know what proportion the Solar Light which we experience, bears to the solar light in its un-diminished state. We do not know whether we have half of it, or a millionth part of it, or almost the whole of it. We have neither test nor *data* for any such statements.

(17.) We do not know what proportion of the Solar Light is absorbed before it reaches our orbit, nor what proportion of it is absorbed by our Atmosphere. We have here, also, neither test nor *data*.

(18.) We do not know anything more of these three proportions with regard to any other planet than we know of them with regard to our own.

(19.) We do not know, therefore, except relatively between planet and planet, what proportion the Solar Light at any other planet, either inside or outside its atmosphere, bears to that which we have either inside or outside ours.

(20.) We only know that the degrees of Light absorbed by equal amounts of Medium—*i.e.*, by equal lengths of a uniform Medium, are equal, since, in nature, the same cause · produces the same effect ; and that in 30 of these equal lengths (or distances from the source) there is 30 times more Light absorbed than in one.

(21.) We know also that, in Nature, there is no diminution of Light possible by Distance alone, without the action of a Medium, there being no distance or space in Nature, without a medium.

(22.) We know that, in a Medium, twice the distance, or length of Medium, does NOT give four times the reduction, or amount lost, nor 30 times the distance 900 times the amount lost. (*See* paragraph 11.)

(23.) We also know that we cannot infer the remainder of an unknown quantity from a series of equal deductions to which it is subjected, even if these were known quantities; —*a fortiori*, then, not when these are unknown, or known only relatively to one another. In other words: when a quantity (of Light or anything else) has undergone 30 equal reductions, in succession, the sum of these reductions is 30 times greater than the first of them; but the quantity remaining after the last reduction is not thereby rendered 30 times less than that remaining after the first reduction. In short "30 times more reduced" does not mean "30 times less."

(24.) According to the Law of the Inverse Squares, the Light from a source at one end of a yard measure, becomes reduced to a millionth part of what it is at the other end, by our merely marking off, upon this yard measure, the 1,000 minute divisions necessary; and stating the light at the end of the last division to be so many times less than that at the end of the first division;—which is absurd.

(25.) According to this same Law of the Inverse Squares, the degree of Light at the end of the second distance from the Source is *always* exactly one quarter of the degree at the beginning of that second distance, whether the distance be a yard or a million of miles;—which also is absurd.

(26.) The Interplanetary Medium, or solar atmosphere, is not uniform. Like our Atmosphere, it is much denser towards the centre. The diminution therefore of the Sun's light, resulting from it, however small this diminution is, becomes thus, by the Sun's attraction, less and less, in proportion to the distance from the Sun at which it takes place;— *i.e.*, this *diminution* itself *diminishes* as the distance from the Sun increases; [and it is estimated by those who advocate the Law of the Inverse Squares that, in all cases connected with attraction or gravitation, such Diminution would itself diminish (or become less and less) at the enormous rate called by them " the square of the Distance."]

(27.) The Solar Light does not therefore pass through nearly 30 times as much medium in its course to Neptune's orbit as in its course to ours; and, in its course to each planet, it only passes through one Atmosphere.

(28.) A given length of Medium does not absorb *more or less* Light according to the amount that enters it. It always absorbs exactly the same amount. If capable of absorbing a given degree or amount of Light, it does so when that amount enters it, whether the whole unreflected light of the Source be that degree only or a degree ten thousand times greater.

(29.) Light can be completely absorbed by the Medium through which it passes, provided the length of the Medium is sufficient in connection with its density, and its density in connection with its length. In no case does Light resist all lengths and all densities.

(30.) The Interplanetary Medium is at least 250 millions of times less dense than the air we breathe. A thirty-fold difference, therefore, between two degrees of Light,—a difference so perceptible in our atmosphere,—would evidently be, in that medium, what the human eye could not discern; but even this infinitesimal amount of difference does not exist between the Sun's light at Neptune's orbit and at ours. (*See* paragraph 26.)

END OF PART IV.

APPENDIX.

APPENDIX.

No. 1.

EXTRACTED FROM PROFESSOR HOGG'S "ELEMENTS OF NATURAL PHILOSOPHY," IN BOHN'S SCIENTIFIC LIBRARY. 1861.

Of Rays and the Divergence- or Spoke-Theory.—" The force of attraction, like many other forces, and like heat and Light, *spreads* in all directions ; that is to say, it *radiates.* The 'lines of force' *pass outwards* from the centre, like the SPOKES from the axle of a wheel. The greater the distance from the centre, the more the lines of force BECOME SEPARATED. The force thus *acts more and more weakly* as it spreads. It spreads in proportion to the square of the distance. The force diminishes in the same ratio ; or, speaking scientifically, it is inversely as the square of the distance from its origin " (p. 16).

" Light is a *radiating* force ; and this law may be proved of Light in a simple manner :—Station at the distance of one yard from a candle-flame a board having a surface of 1 square foot; at another yard distance, place a board of a surface of 4 square feet ; at another yard further off again, a board of 9 square feet superficies. The shadow cast by the first board will exactly cover the second, and that by the second will cover the third. The distance of the first board·

being 1, that of the second is 2, the square of which is 4, the number of feet of surface; that of the third is 3, the square of which is 9, in like manner. THIS SHOWS THE PROGRESSIVE SEPARATION OF THE RAYS. We may next demonstrate that the intensity of the light diminishes in precisely the same proportion " (p. 17).

" The rays of Light diminish in intensity (or compactness and density) according to Distance, in correspondence with the law that governs many forces and radiating influences in general. Light diminishes as the square of the distance increases, in the proportion of 1, 4, 16, &c., to the distances 1, 2, 3, 4, &c."

On the Spreading of Light from Ray to Ray.—" Its VOLUME at the same time increases; for if a candle be lighted on a dark night and placed in an open prominent position, the light will fill a sphere of a mile in diameter " (p. 281).

The number of distinct rays must baffle all attempts at calculation; for were the entire space covered with eyes, all would receive a portion and become sensible of its presence " (p. 281).

On the Effects of Medium.—" The finest glass manufactured is not perfectly transparent " (p. 283).

" Transparent bodies do not transmit all the light that enters them. *Air arrests a considerable portion of light*, as it passes through it; and (the clearest) water does not allow it to penetrate a depth beyond seven feet without absorbing one-half of its quantity. Practised divers know that in the clearest water they soon find darkness as they descend, and that the bottom of the ocean is a world without light" (p. 278).

Professor Hogg, although advocating in some passages the Spoke-theory, speaks, we see, in others, of the converging cones also; *i.e.*, of there being a ray proceeding from each point of the disc to each point of minutest space (*e.g.*, of lens or retina), recognizing the universality and, therefore, convergence of all rays, thence resulting. He speaks as follows of the way in which the diverging cones of rays, as well as the converging ones, become nearly cylinders :—

" It is to be observed that the further an object (radiating light) is from the lens (or eye), the more nearly do the rays darting from it towards the lens (or eye) approach to being parallel. If the distance of the radiant *point* be very great, they really are so nearly parallel that a very nice test is required to detect the non-accordance (or convergence). Rays, for instance, coming to the earth from the sun, do not diverge (at their point on the sun) the millionth of an inch in (*i.e.* for) a thousand miles.   Hence, when we wish to make experiments with parallel rays, we take those of the sun " (p. 300).

No. 2.

TRANSLATED FROM PROFESSOR LOMMEL'S " NATURE OF LIGHT" (NATUR DES LICHTES. ERLANGEN, 1875).

" Light proceeding from a luminous body whilst traversing a homogeneous medium is propagated in every direction in straight lines, which are called Rays of Light" (p. 14).

At p. 20, he speaks again of " *the rectilinear course of the rays of Light.*"

Then at p. 22, he supposes a luminous point and two opaque screens placed parallel in line before it in a medium ; and says: " If the second screen be just twice as distant from the source of light as the first, the area of the shadow will be *four times as large* as the screen which throws the shadow. If the latter be removed, the same number of rays, which was previously received by it and illuminated its surface, is now distributed over an area of four times the size; a given portion of the second screen receives, consequently, 4 times less light than a corresponding portion of the first screen, and will be therefore proportionately less strongly illuminated. The source of Light thus gives, at double the distance, only the fourth part of the illumination which it can give at unity. If the second screen be at 3, 4, 5 times the distance of the first from the source of Light, the shadow falling upon it will be 9, 16, 25 times larger than the shadow-throwing screen, and will, according to its distance, be 9, 16, 25 times less brilliantly illuminated. . . . We thus acquire a knowledge of the law, that the amount of illumination diminishes in proportion to the square of the distance from the *source* of illumination" (pp. 22-3).

Of the Medium Photometer.—Professor Lommel then proceeds "to demonstrate the truth of this law," as he expresses it, "by experiment" *in a Medium:*

"A sheet of white paper is stretched on a frame, supported on a stand, in the centre of which (paper) is a spot of oil, made with stearine. The grease spot allows more light to pass through it, and consequently reflects less than the unstained part of the paper. If therefore the paper be illuminated more strongly from behind, it appears bright on a dark ground. On the other hand, it appears dark upon a bright ground if it be more strongly illuminated on the front surface; whilst with equal illumination on both sides, the spot becomes invisible, since it can then appear neither darker nor lighter than the adjoining paper. The flame of a candle is now placed upon one side of the screen, whilst four such flames are placed upon the other side, and the screen is removed to such a distance from these that the spot is no longer visible. This will be found to occur when the distance of the quadruple flame from the screen on the one side is double that of the single flame on the other side. This experiment, in which a source of light four times as strong as another gives the same illumination at double the distance, corroborates the law above laid down.

"This law being *admitted*, the same apparatus may be employed as a means of comparing the brilliancy of two sources of light. If, for example, the flame of a candle be placed in front of and a gas flame behind a paper screen, and this be moved till the grease spot disappears, the illuminating power of the two sources will be as the squares of their distances from the screen. The apparatus employed for the determination of the illuminating powers of different sources of light are termed Photometers."

The Professor then describes another of these Medium Photometers, — Photometers acting in and through a Medium,—in order " to demonstrate" what happens where there is no Medium :—

" An opaque rod, about the size of a lead pencil, stands in front of a white paper screen. The two lights to be compared both cause a shadow of the pencil, and each light illuminates the shadow cast by the other. If either light is removed to such a distance that the two shadows appear of equal depth, the brilliancy of the two lights will be as the squares of their distances from the screen" (pp. 23-5).

Of Converging Cones.—" The *apparent size* of an object is determined by the angle which the rays of light, passing from its outermost points to the eye, form with one another, —the so-called visual angle. The same body is seen under a smaller visual angle, and of correspondingly smaller size the further it is removed from our eyes ; and two bodies of different size appear under the same visual angle if their distances are inversely as their diameter" (pp. 17, 18).

After speaking of two of their theories, he says further of the converging cones or the Universality of Radiation:

" Whilst on the older theory, a direct propagation along a single straight line, *i.e.*, the possibility of an *isolated ray* of light, was accepted ; on the other theory, in view of the action which every particle of æther (the *luminous* æther) exercises upon the adjoining ones, *the existence of an isolated ray of light is inconceivable.* Nevertheless, ' a ray of light' may be conceived as the expression of the *direction* in which (the light) is propagated " (p. 232).

" The statements formerly made on the supposition of *individual rays* of light, lose none of their force through the different conception now gained" (p. 233).

No. 3.

EXTRACTED FROM PROFESSOR TYNDALL'S "NOTES ON LIGHT."

"Light moves in straight lines."

"Light diminishes in intensity as we recede from the source of light." (He means as the upright area illuminated is farther from the source. He does not mean that *we* make the Light more or less intense by our position.) "If the luminous source be a point, the intensity diminishes as the square of the distance increases. Calling the quantity of light falling upon a given (upright) surface at the distance of a foot or a yard, 1, the quantity falling on it at a distance of 2 feet or 2 yards is $\frac{1}{4}$; at a distance of 3 feet or 3 yards it is $\frac{1}{9}$; at a distance of 10 feet or yards it would be $\frac{1}{100}$, and so on. This is the meaning of the Law of inverse squares as applied to light" (p. 3).

He then, in his "experimental illustrations" with a medium, speaks of this diminution in nearly the same terms as Professor Lommel. He supposes 4 equal distances taken in a line from the luminous body, and says that the Light at the end of the fourth distance "is *diffused over* 16 times the area" over which it was diffused at the end of the first distance; and adds: "It is therefore *diluted* to $\frac{1}{16}$ of its former intensity; that is to say, by augmenting the distance 4-fold, we diminish the Light 16-fold."

He next speaks of the third distance, and says that the Light at the end of it "is diffused over 9 times the surface" over which it was diffused at the end of the first distance; adding: "It is therefore *reduced* to $\frac{1}{9}$ of its intensity. That is to say, trebling the distance from the source of Light, we diminish the Light 9-fold."

He then speaks in similar terms of the light at the end of the second distance, saying that it is "*diluted*" to a fourth of the intensity existing at the end of the first distance, by

being spread out or diffused over 4 times a larger area."
(*See* page 4 of the "Notes on Light.")

On the Perspective Theory for this diminution of Light by Distance, or the Diminution of the Apparent Area, he says : " As we retreat from a light (he means a luminous *body*), its image upon the retina becomes smaller, and it is easy to prove that the diminution (of the area) follows the law of the inverse squares ;—that at a double distance the area of the retinal image is reduced to one-fourth ; at a treble distance, to one-ninth, and so on. The concentration of Light accompanying this decrease of Magnitude exactly atones for the diminution due to distance" (p. 5).

At p. 71, he writes : " Sir William Thomson has attempted to calculate the mechanical value of *a cubic mile* of sunlight." And again : " What then is the thing moved in the case of our *cubic mile of sunlight ?*" " What are the effects which *this cubic mile of Light and Heat* can produce ? "

He speaks thus of the illustration which a MEDIUM affords of the " Inverse Squares" and the " Square of the distance " :—

" Place an upright rod in front of a white screen and a candle flame at some distance behind the rod, the rod casts a shadow upon the screen. Place a second flame by the side of the first ; a second shadow is cast, and it is easy to arrange, &c."

" Measure the distances of the two lights (he means *sources*) from the screen, and *square* these distances. The two squares will express the relative illuminating powers of the two lights. Supposing one distance to be 3 feet and the other 5, the relative illuminating powers are as 9 to 25 " (p. 5).

On " the Law of the Inverse Squares," as a mere opinion, he further says in a letter of July, 1874 : " The subject of which you write is one regarding which OUR MINDS ARE ALREADY MADE UP ; and I see so little hope of any *useful* result coming out of the discussion of it, that I fear with reluctance I cannot give you any hope of my being able to enter upon it."

In his American Lectures on Light (First edition), Professor Tyndall says on the Spoke-Theory or Radiation: "The ancients were aware of the rectilineal propagation of light. . . . Possibly the terms 'ray' and 'beam' may have been suggested by those straight SPOKES of light which, in certain states of the atmosphere, dart from the sun at his rising and his setting" (p. 9).

"It is confusion and stagnation, rather than error, that we ought to avoid" (p. 80).

No. 4.

TRANSLATED FROM PROFESSOR AUGUST KUNZEK'S TREATISE ON "LIGHT" (DIE LEHRE VOM LICHTE. VIENNA).

"We see that the rays of Light separate from one another as they leave each point of the source; that is, they become divergent; and only those rays which from a very distant source fall upon a minute surface, can be regarded or treated as parallel rays" (p. 21 of the original).

" *On the Diminution of Intensity in Consequence of Distance from the Source:*—The emanation of Light in straight lines from the luminous body, and the *divergence* of these straight lines *from one another* which *necessarily results* from this circumstance, is the REASON why the degree of illumination upon any surface must diminish as the distance from the source increases" (p. 31).

On the Diminution in Consequence of the Medium, he says:— "When a ray of Light passes through a medium, it becomes more and more diminished in intensity in proportion to the distance which it traverses in that medium.

"The law of this diminution is as follows:—If we suppose a uniform medium, *i.e.*, one which is of the same nature and density throughout, then each equal length of this medium will absorb, not the same quantity of Light, but exactly the same proportion of the Light which enters it;—*i.e.*, if the degree of Light which enters the first of these equal lengths is reduced by a given fraction (say $\frac{1}{6}$ of it), then the degree which enters the second and third equal length will also be reduced by precisely the same fraction of its own amount (*i.e.*, by $\frac{1}{6}$ of itself).

" Light, therefore, in passing through a medium, decreases in a geometrical progression, while the length of the medium (*i.e.*, the distance from the source) increases in an arithmetical one."

No. 5.

Sir William Thomson, Professor of Physics in the University of Glasgow, in a letter in which he declines to give his sanction to any discussion respecting the truth of the Inverse Squares as applied to Light, states, as his reason for objecting to this discussion, that he is " QUITE CONVINCED of the truth of the ordinary scientific view regarding the Solar Illumination."

The Professor also makes the following admirable remark in the Preface to his " Elements (*Mathematical*) of Natural Philosophy," 1872 :—

" *Simplification of Modes of Proof* is not merely an indication of advance in our knowledge of a subject, but is also the surest guarantee of *readiness for future progress.*"

See No. 3 for another remark of this Professor's.

No. 6.

When speaking of Bouguer's Treatise, *Gradation de la Lumière*, 1729,—the earliest Treatise on the diminution of Light by Distance (*see* No. 12)—Sir John Leslie, Professor of Physics in the University of Edinburgh, remarks :—" He there sets out from the OBVIOUS principle that Light, darting in straight lines, *must* become *dilated* or attenuated in the ratio of the square of the distance from the radiant source." (*See* Leslie's Dissertation, "Light," in the Encyclopædia Britannica).

No. 7.

On the Perspective Theory for the Diminution of Light (the subject also of an important paragraph in No. 3 of these Extracts), I find the following passage in a letter from Mr. Proctor. After observing that the apparent brightness of a disc does not diminish with distance, he adds:—" But its apparent size, and THEREFORE the total quantity of light, received from it, does diminish,—as the square of the distance increases."

On the Law of the Inverse Squares and the Solar Illumination of the System, supposed thence to result, Mr. Proctor also writes:—" I must frankly say,—though I have no wish to discuss the matter,—that all the observed phenomena appear to me to accord with the usual Theory of the distribution of Light; and that I have not yet heard of any reason—theoretical or otherwise—justifying any question of its validity."

No. 8.

A President of the British Association says, respecting one of my printed statements :—

" I herewith return Dr. Simon's paper, which I have read. It seems to me, however, that he has involved himself in difficulties of his own seeking, for the Law in question simply states the fact that the amount of Light falling upon a given area diminishes inversely as the square of the distance. . . . This Law is in fact NOTHING BUT THE GEOMETRICAL LAW OF AREAS stated optically."

No. 9.

Sir David Brewster, in his "Optics," says :—

On the Divergence Theory.—(4.) "Light is emitted from every visible point of a luminous or of an illuminated body, and *in every direction* in which the point is visible. If we look at the flame of a candle, or at a sheet of white paper, and magnify them ever so much, we shall not observe any points destitute of light."

(5.) "Light moves in straight lines, and consists of *separate and independent* parts called rays of light."

On the Law of Absorption.—(85.) "Some idea may be formed of the law according to which a body absorbs light, by supposing it to consist of a given number of equally thin plates, at the refracting surfaces of which there is no light lost by reflection. If the first plate has the power of absorbing $\frac{1}{10}$ of the light which enters it, or 100 rays out of 1,000; then $\frac{9}{10}$ of the original light, or 900 rays, will fall upon the second plate; and $\frac{1}{10}$ of these, or 90 being absorbed, 810 will fall upon the third plate, and so on. Hence it is obvious that the quantity of light absorbed by any number of films is equal to the light transmitted through one film multiplied as often into itself as there are films. Thus, since 1,000 rays are transmitted by one film, $\frac{9}{10} \times \frac{9}{10} \times \frac{9}{10}$ equal to $\frac{729}{1000}$ or 729 rays will be the quantity transmitted by three films; and therefore the quantity absorbed will be 271 rays."

No. 10.

Sir John Herschel, in the article " Optics " in the Encyclopædia Metropolitana, writes :—

Of the Divergence Theory and the Inverse Squares.—" In a
free medium " (the technical phrase for no medium at all)
" the force and intensity of light which propagates itself
in rays emanating from the same point, or which concur
in the same point, are inversely as the squares of the
distances from that point.

" For the deviations from each other of two rays of light
which proceed from the same point, are always proportional
to their distances from that point (since those deviations
form parallel bases of isosceles triangles, of which the two
rays are the sides). Suppose, therefore, that having intercepted a certain number of rays by a plane, posited at a
certain distance from the radiant point, we remove this
plane to a double distance, then to a triple, to a quadruple
distance, and so on; the deviations of the rays from each
other will be as the numbers 1, 2, 3, 4, &c., and each dimension of the base of each luminous pyramid which is thus
formed in succession, will be in the same ratio. Consequently, the surface of those bases will be as the numbers
1, 4, 9, 16, &c. So that the same number of rays are found
distributed successively over surfaces which are respectively
as the squares of the distances from the radiant point, or
point of concourse, and therefore the intensity of light that
they excite will diminish in the same proportion " (p. 12).

Of the Medium, Herschel says :—" The simplest hypothesis
we can form of the extinction of a beam of homogeneous
light in passing through a homogeneous medium, is, that for
every equal thickness of the medium passed through, an
equal aliquot part of the rays which, up to that depth had
escaped absorption, is extinguished. Thus, if 1,000 red
rays fall on and enter into a certain green glass, and if 100
be extinguished in traversing the first tenth of an inch,

there will remain 900 which have penetrated so far; and of these one-tenth, or 90, will be extinguished in the next tenth of an inch, leaving 810, out of which again a tenth, or 81, will be extinguished in traversing the third tenth, leaving 729, and so on. In other words, the quantity unabsorbed, after the beam has traversed any thickness of the medium, will diminish in geometrical progression, as the space traversed increases in arithmetical.

"It is evident from this, that, strictly speaking, total extinction can never take place by any finite thickness of the medium" (pp. 488 & 489).

No. 11.

J. H. Lambert, in a Treatise "On the Passage of Light through the Medium" (Sur la route de la Lumière par les airs), published at The Hague in 1759, writes as follows:—

"The other department of Optics" (not Reflection nor Refraction) "which I chiefly now propose to treat of, is called Photometry. The subject here whose laws are stated is the Intensity of Light, the closeness of its rays,—in other words, its illuminating power, the increase and decrease which under any circumstance it undergoes, &c. If the first portion of this science (the Reflection and Refraction of Light) has been of vast importance to science and to mankind, Photometry contributes a much larger share to the same result.

"Any one who seeks to establish a Theory of Light will find that it is not enough for us to know that Light is reflected and refracted according to certain laws. He will also easily recognize the importance of being able to know, as deductions or inferences, the degrees of Light connected with these changes of reflection and refraction, and to know this through experimental illustrations.

"And this branch of Optics is by no means virgin soil,—

by no means an unknown land. Learned men, highly distinguished, have already worked at these questions before I have done so. M. Bouguer has published an excellent Treatise on this diminution of Light by distance. He investigates the laws and causes of this diminution, &c., &c." (pp. 4 and 5 of the original).

The following is the French text of the foregoing:—

L'autre partie de l'Optique dont j'ai principalement dessein de parler, c'est la Photométrie. Elle s'occupe de l'éclat de la lumière, de sa densité, de sa force illuminante, etc., des accroissements et diminutions qu'elle souffre dans tous les cas. Si la première partie de l'Optique a été d'un secours infini, etc., la Photométrie y contribue infiniment plus. Qui veut imaginer une théorie de la lumière, il ne lui suffira pas de savoir qu'elle se refléchit et se brise suivant une certaine loi ; mais il lui importera d'en pouvoir déduire la quantité de l'une et de l'autre conformément aux expériences.

La Photométrie n'est pas un pays entièrement inculte. Des savans fort célèbres y ont travaillé. M. Bouguer en a donné un très bel Essai sur la Gradation de la Lumière, etc. Il en cherche l'affoiblissement, &c. (pp. 4, 5).

No. 12.

Translated from the work on the two Diminutions which Light undergoes in its passage from the Source to the Object, entitled " Sur la Gradation de la Lumière." Paris, 1729,—a work by the distinguished French Physicist Bouguer, who was the first to propound publicly the doctrine of the Inverse Squares, as well as the Law for the diminution in a medium :—

Of the Inverse Squares.

" FIRST SECTION.—*Modes of measuring the intensity (amount, degree, or quantity) of Light.*—One of the more important peculiarities of Light is that of being *divided* naturally into *separate* rays which all go off from the luminous point, in *straight* lines, *independently of one another*, and even *in quite different directions*, which rays thus form a kind of pyramid (of dark space between them), the apex of the pyramid being in this case the source of Light (as it is thus also source, we see, of the resultant darkness between the rays).

From this peculiarity it follows that Light becomes weaker and weaker in proportion *as the upright area upon which it falls is more and more remote from the source of light.* If the retina or other upright area illuminated is at first quite near, and is afterwards removed fifty or sixty steps farther from the source, then several rays, which fall on it at the first distance, will now fall wide of it, and so miss it altogether; because these rays, proceeding in a perfectly *straight* line from the source, *separate continually more and more from each other* (leaving unilluminated spaces between them); so that the rays which were very dense and *close together* at first, are at last *dispersed wide apart over* a large extent of surface, while the force (amount or intensity) of the Light must therefore be, inversely, as the squares of the distance from the luminous body: and this for the very substantial reason that the *objects* illuminated,—the upright surfaces, areas, or spaces over which the rays have to spread,—become greater and greater at these greater distances, in the direct ratio of these very squares.

"To make this fact more obvious, although it is well known to all those initiated in Optical research, we shall suppose that, after an object or area has received the light of a lamp at a certain distance, this area or object is afterwards placed *three* times farther from the lamp than it was at first. Then, the rays which dart off on that side, and at that point of the lamp, will now occupy *nine* times a larger object or upright area than before; because this remarkable peculiarity of *all* rays,—their divergence from one another on leaving the luminous body,—will compel them to fall upon an object (upright surface or space), *three* times higher and three times wider than at first. Since, however, the same *number* of rays are now to be spread over an object or empty space *nine* times more extensive, nine times fewer of them will fall upon each spot of this more extended object, or more extended empty space, presented to them; and the Light will therefore be nine times weaker in amount upon each spot of this object (although the same in amount upon the whole of it).

" If in like manner the area or object is placed at a distance *ten* times greater than at first, then the rays, which always go on separating from one another with greater and greater intervals (of darkness), will necessarily occupy a *hundred* times a larger object (or a hundred times more empty space), and each spot of that object or that·space will consequently be a hundred times less illuminated. In short, the rays, as we have said, are always forming a pyramid (or divergent cone of darkness) whose apex is, as it were, the luminous body, and whose base is this enlarged object or upright space or surface, upon which all these separated rays thus fall.

" Now since this upright base or object must always *be rendered* (by its position with regard to the apex) more extensive in the exact ratio known as the square of the distance from the beginning of the supposed cone or pyramid (of darkness), and since this dark object or base receives, nevertheless, over its whole extent, only the same number of rays, although it is so much increased in size, it is evident that it will receive fewer of these rays upon each 'unity of surface,' and that the amount of the Light (its degree, force, or intensity) will therefore be less and less, in this relative sense (*i.e.*, in proportion to the space it illuminates) as the square of the distance is greater and greater.

*   *   *   *   *   *   *   *

" All that is here necessary is, that the space between the enlarged object (the upright surface or space) and the luminous point should not be too great; in order that the Light should be *perceptibly* diminished, in this relative sense, and by this divergence only, not by any medium,—*i.e.*, only by the increasing spaces between the rays, and not by the amount of obstruction which the atoms of the medium would present in the course of a longer passage. With that proviso, it is always easy to alter the amount (force or intensity) of the Light to any extent that we desire. We have only to let it fall upon dark objects or areas geometrically proportioned to the different distances from the apex at which we judge of it. Nothing can be simpler or more familiar than such a process. Nor shall we now employ any other for estimating the strength of Light, and for determining various other problems which people have either hitherto in vain attempted to solve, or have never yet taken it into their heads to study."

Of Diminution in the Medium.

In Section II., Chap. 1 (with the heading " *On the Way in which Light passes through a Medium*"), Bouguer explains how Light *may be* supposed to pass through the atoms of all media; and then in Chapter 2, proceeds thus to state the law for this second kind of diminution :—

"As we have explained (in Chap. 1) the way in which Light passes through all kinds of medium, so we must now investigate the law by which its diminution there takes place.

" The idea which would naturally occur to us upon this subject is, that if we suppose a medium divided into several parallel layers of the same thickness (*i.e.*, width) and density, each of these layers absorbs or intercepts precisely the same number of rays,—precisely the same amount of Light ;—so that the Light, encountering exactly the same amount of diminution in its passage through each layer, would simply diminish in what is called arithmetical progression (*i.e.*, 30 times more distance, or thickness, would give 30 times more reduction or diminution, but not 30 times less Light). This is the natural view of matters.

" To ascertain, however, the truth or fallacy of this opinion, I once made a light equal to 32 candles pass perpendicularly through two pieces of glass, and the light was thereby reduced to one-half, being then only equal to the light of 16 candles. If, now, two other similar pieces of glass (two other layers of medium, *i.e.*, lengths or distances from the source) had produced a similar effect,—an equal diminution of light,—it is evident that all the rays would have been absorbed or intercepted,—all the light would have become extinct; and *a fortiori*, 8 or 10 pieces of glass (*i.e.*, lengths of medium, or distances in the medium) would have presented a medium utterly impenetrable to the 32 candles. Nevertheless, when I added only two more pieces of glass to the two first employed, they were very far indeed from absorbing all the light of the 32 candles. The light was still very considerable; and when I made it pass through ten pieces of similar glass, it was, even then, manifestly as strong as the light of a single candle.

" But I have done enough in having thus drawn the attention of studious people to this matter. They will no doubt easily see that a second layer (or distance) of the medium could absorb or intercept precisely the same number of rays as the first, only upon one condition; viz., that *precisely the same number of rays should fall upon this second layer, as fell upon the first.* But since this is impossible,—since there probably does not reach this second distance or layer of the Medium more than a third or fourth part of all the rays that entered the first layer, the rest having been already absorbed by that first layer, it is clear that this second space or distance, or layer of medium, must also therefore necessarily absorb three or four times less Light,—three or four times fewer rays,—than were absorbed by the first distance. Equal layers (or equal distances) THEREFORE cannot absorb equal quantities of Light, but only equal proportions. If, for instance, a given distance or layer absorbs half the Light, the next equal measure of the medium, in all respects equal to and similar to the first, will not absorb the whole of the other half, but only the half of it,—the half of that other half ;—and will therefore reduce it to one quarter of the original Light; and as all the other layers will absorb similar proportions of the Light which enters them, it is evident that Light in a medium is always diminished in this way, technically called a geometrical progression."

Three detached sentences of this very frank writer may here be usefully cited, without citing, however, the arguments which they introduce :—

" It follows, from what I have said, that if we suppose the luminary to be extremely remote, in order that the rays should *seem* parallel, and that the Light should be diminished *only* by the Medium and the atoms of the Medium, without this insensible divergence of the rays being able to produce, at the same time, any sensible effect in its diminution,—then I repeat it follows," &c.

Section IV. begins thus :—

" We above supposed the source to be extremely remote, and that the rays proceeding from it were apparently parallel. Let us now suppose the luminary not to be so remote, and let us observe the effect which necessarily results from the divergence of its rays. In this case, the amount (degree or intensity) of the light, is sensibly subjected to *two different kinds of diminution at the same time ;*—one from the density of the medium and the absorption which this always occasions, and the other from the divergence of the unabsorbed rays; for the rays which succeed in traversing the medium and are not absorbed, go on always separating more and more from one another, (with a constant and considerable increase of the dark intervals between them); the result of which is that fewer of these rays reach the same area at the greater distance. If now this second diminution of Light, which occurs in the ratio explained as the Inverse Square of the distance, be added to that other diminution which results from the absorption effected by the medium, we shall then see, &c., &c."

And again : " Although the rays proceeding from a lamp
form a cone of dark space by their divergence, and go on
always becoming more and more distant from one another,
we must not suppose that they are on that account more
liable to be absorbed, &c."

In consequence of the importance of this writer in connection with this subject, both on account of his being the
first to propound publicly the two theories in question,
from whom clearly all succeeding writers have derived
their reasonings and statements, and on account of a frank-
ness and distinctness in what he has written, which we
rarely find in other writers respecting either of these
theories, I subjoin the original of the foregoing extracts,
and the original also of two or three other passages from his
writings, which it did not seem so necessary to translate :—

Bouguer. Sur la gradation de la lumière.

Section I. *Méthodes de mesurer la force de la lumière.*

Chapitre 1. Une des principales propriétés de la lumière
est d'être distribuée en plusieurs rayons, qui avancent tous,
chacun *à part*, en ligne droite, et qui forment une espèce de
pyramide dont le corps lumineux est le sommet. Cette pro-
priété est cause que la lumière devient plus foible, à mesure
qu'on la reçoit à une plus grande distance du corps lumineux.
Si on en est d'abord très-proche et qu'on s'en éloigne ensuite
de 50 ou 60 pas, plusieurs rayons qui entraient dans l'œil
en tomberaient fort loin; parcequ'en avançant en ligne
droite, ils s'écartent tous de plus en plus les uns des autres.
De cette sorte, les rayons qui étaient très-denses ou très-
serrés, se trouvent dispersés dans une grande étendue et la
force de la lumière doit suivre la raison inverse des carrés
de la distance au corps lumineux; parceque les espaces
dans lesquels les rayons se trouvent répandus augmentent
en raison directe de ces mêmes carrés.

Pour rendre cette vérité plus sensible quoiqu'elle soit
réconnue de tous les lecteurs qui sont initiés dans l'Optique,
nous supposerons qu'après avoir reçu la lumière d'un
flambeau à une certaine distance, nous nous en mettions à

une distance trois fois plus grande. Les rayons du flambeau qui viennent vers nous, occuperont dans le second cas une étendue neuf fois plus grande que dans le premier ; car leur divergence sera cause qu'ils tomberont sur un espace qui aura trois fois plus de largeur et trois fois plus de hauteur. Mais puisque la même quantité de rayons se distribue dans une étendue neuf fois plus grande, il en tombera neuf fois moins en chaque endroit, et la lumière sera par conséquent neuf fois plus foible. Si on se met pareillement à une distance dix fois plus grande, les rayons qui vont toujours en s'éloignant les uns des autres, occuperont un espace cent fois plus grand, et chaque endroit sera cent fois moins éclairé. En un mot, les rayons forment toujours, comme nous l'avons dit, une pyramide dont le corps lumineux est le sommet, et dont la base est la surface sur laquelle on reçoit la lumière. Or comme cette base augmente précisément en même raison que le carré de la distance au corps lumineux, et qu'elle ne reçoit cependant toujours dans toute son étendue que le même nombre de rayons, il est sensible qu'elle en recevra moins dans chaque de ses points, et que la force de la lumière sera précisément plus petite en même raison, que le carré de la distance sera plus grand.

Ce que nous disons ici, peut s'appliquer aussi aux foyers des verres ardens ou des miroirs concaves ; car après que les rayons se sont croisés dans ces points, ils deviennent divergens et ils occupent des espaces qui augmentent en même raison que les carrés des distances au foyer. Mais la lumière, répandue de cette sorte,—dans des espaces qui sont plus grands,—doit être aussi plus foible précisément en même raison.

Il suffit en tout cela que les distances du corps on du point lumineux ne soient pas excessives, afin que la lumière ne diminue *sensiblement* que par sa seule divergence, et non pas par le rencontre des parties grossières de l'air qui pourraient intercepter plusieurs rayons dans un plus long trajet. Cela supposé, nous pourrons toujours faire changer très-aisément la force de la lumière en quelle proportion nous voudrons ; nous n'aurons qu'à la recevoir à différentes distances du corps lumineux, ou bien à différentes distances du foyer dans lequel nous l'aurons réunie, en nous servant d'un verre convexe. On ne peut rien concevoir de plus simple ni de plus connu que ce moyen. Cependant nous n'en employerons point d'autres pour mesurer la force de la lumière et pour déterminer différentes choses qu'on a tenté inutilement jusqu'ici de découvrir, ou qu'on ne s'est point encore avisé de chercher.

Section II. *De la manière dont la lumière passe au travers des corps.*

Chapitre 2. Puisque nous avons expliqué la manière dont la lumière passe au travers des corps diaphanes, il nous faut examiner à présent selon quelle loi elle diminue dans ce passage. La première pensée qui se présente sur ce sujet, c'est que si on conçoit un corps diaphane, divisé en quantité de couches parallèles de même épaisseur, toutes ces couches intercepteront le même nombre de rayons ; de sorte que la lumière recevant dans le passage de chaque tranche une diminution toujours exactement égale, elle décroîtrait en progression arithmetique, ou en suivant le rapport des ordonnées d'un triangle.

Pour réconaître la vérité ou la fausseté de ce sentiment, j'ai fait une fois passer perpendiculairement au travers de deux morceaux de verre une lumière qui était égale à celle de 32 chandelles, et elle ne se trouva ensuite deux fois plus foible ; car elle ne se trouva plus égale qu' à la lumière de 16 chandelles. Or si une autre épaisseur de deux morceaux de verre eut produit un égal affoiblissement, il est évident que tous les rayons eussent été interrompus ; et à plus forte raison, 8 ou 10 morceaux de verre eussent formé une épaisseur tout-à-fait impénétrable à la lumière. Cependant ayant ajouté 2 morceaux aux deux premiers il s'en fallut beaucoup qu' ils ne formassent un corps absolument opaque ; la lumière se trouva encore très-vive ; et lorsque je la fis passer au travers de dix morceaux, elle était encore sensiblement aussi forte que celle d'une chandelle.

Mais sans doute qu'il suffit d'y avoir fait penser les lecteurs, et qu'ils voyeut bien que pour qu'une seconde épaisseur interceptât précisément le même nombre de rayons que la première, il faudrait qu'il se presentât aussi précisément le même nombre de rayons pour la traverser. Mais puisqu'il ne parvient peutêtre à cette tranche que le tiers ou le quart du nombre total des rayons parceque tous les autres ont dèjà été interrompus, il est certain que cette tranche doit intercepter aussi trois ou quatre fois moins de rayons que la première. Ainsi les tranches égales ne doivent pas détruire des quantités égales, mais seulement des quantités proportionelles. C'ést-à-dire, que si une certaine épaisseur intercepte la moitié de la lumière, l'autre épaisseur qui suivra le première, et qui lui sera égale, n'interceptal pas toute l'autre moitié, mais seulement la moitié de cette moitié, et la réduira par conséquent au quart; et toutes les autres tranches détruisant de semblables parties, il est sensible que la lumière diminue toujours en progression géométrique.

Il est clair aussi que ce que nous disons doit être également vrai, de quelque manière que la lumière se transmette au travers des corps transparents. Car, supposons que les rayons ne puissent passer que par les pores, et qu'il y en ait une si grande quantité que les parties solides ne fassent que la centiéme partie du volume extérieur que le corps parait occuper; si on conçoit ce corps divisé en un nombre presqu' infini de tranches, dont l'épaisseur soit égale au diamètre de ces petites parties; la première tranche n'interceptera que la centième partie des rayons, et de 100,000, il y en aura 99,000, qui parviendront à la seconde tranche. Et comme il y aura aussi 100 fois plus de pores dans la seconde tranche que de parties solides, à cause de l'homogénéité du corps, il est clair que la multitude des rayons diminuera encore de la centième partie, en traversant la seconde tranche, et qu'elle se réduira à 98,010. Or toutes les autres tranches produiront un semblable effet; elles feront toujours diminuer la lumière de la centième partie et ainsi la progression géométrique sera toujours exactement observée. D'un autre côté, si les petites parties solides dont les corps sont composés servent souvent elles-mêmes à transmettre la lumière, comme nous avons taché de le prouver, ce sera encore la même chose. Car on peut considérer comme des pores ces grains de matière qui transmettent les rayons et on peut fort bien ne faire attention qu'aux autres grains qui détournent ou qui affaiblissent la lumière, et comme il s'en trouve toujours dans chaque tranche un égal nombre de ces derniers, il est sensible qu'ils feront toujours décroitre la lumière d'une semblable partie, ou d'une partie proportionelle.

Il suit de-là, que si nous supposons que le corps lumineux est infiniment loin, afin que ses rayons soient sensiblement parallèles et que la lumière ne diminue que par la seule interposition du corps transparent sans que la divergence des rayons y ait aucun part ;—il suit, dis-je, de-là que les forces qu'a la lumière, après avoir traversé différentes épaisseurs peuvent être représentés par les ordonnées d'une triangle logarithmique qui a pour axe l'épaisseur du corps.

Section IV. begins thus :—

Nous avons supposé ci-devant que le corps lumineux était à une distance, comme infinie, et que ses rayons étaient sensiblement parallèles. Nous allons maintenant supposer que le corps lumineux n'est plus si éloigné, et faire attention à l'effet que doit causer la divergence de ses rayons. Dans ce cas la vivacité de la lumière est sujette à recevoir deux diminutions ; l'une par le défaut de transparence du milieu, lequel intercepte toujours quelques rayons ; et l'autre parceque les rayons qui restent et qui ne sont point interrompus vont toujours en s'éloignant les uns des autres, et occupent continuellement de plus grands espaces ; ce qui fait qu'il en tombe moins en chaque endroit. Or si on combine cette dernière diminution qui suit (comme nous l'avons vu ci-devant) la raison inverse des carrés des distances, avec l'autre diminution causée par le défaut de transparence, on verra, etc., etc.

Si les rayons qui sortent d'un flambeau forment un cone par leur divergence et vont toujours en s'éloignant les uns des autres, ils ne doivent pas être pour cela plus sujets à être interrompus, etc., etc.

No. 13.

Translated from Fischer's "Natural Philosophy" published with Biot's Notes.

Of the Two Kinds of Diminution.—" But the Light which proceeds from a body loses its intensity by diffusing itself, *since it is spread over a space the more extended the farther it travels from the body.* By means of well-known geometrical theories it may be demonstrated *that the intensity of light is inversely proportional to the square of the distance,* on the supposition that it is not diminished by any other cause *except divergence of the rays*" (p. 495).

Translated from a note by Biot on the above :—

"Let A be a radiating point, and at the distance A B suppose a geometrical plane B C D perpendicular to A B. It is obvious that the light (*i.e.* the delineating rays), falling from A upon B C D must have the form of a pyramid (enclosing a dark space), whose vertex is A, and whose base is B C D. If we prolong this pyramid indefinitely, and at some distance A E, taken at pleasure, cut it by a plane E F G parallel to the former B C D, it is obvious that there must be as much light in one of these planes as in the other. But B C D is smaller than E F G, and consequently the light must be more concentrated in the former, in the ratio of the two surfaces."

No. 14.

In a letter from one of the ablest and most enlightened, as well as most distinguished and amiable of our English Professors, I find the following :—

"The reason why I did not think your paper, as far as I could judge of it, suitable for presentation to the Royal Society, was that the law of illumination, varying as the Inverse Square of the distance seemed to me so well established that, weighing the known arguments for, against the unknown arguments against, I felt a full expectation that when those unknown arguments came to be examined, a flaw would be discovered in them. It was not, as you suppose, a disinclination to disturb the popular belief. All science requires is a fair field and no favour.—May, 1878."

No. 15.

EXTRACTED FROM DR. ARNOTT'S "ELEMENTS OF PHYSICS." —Ed. 1876.

Of the Spreading of Light.—"The conditions on which depends the intensity of gravitation,—as well as of light, magnetism, sound, or any other influence spreading uniformly from a centre,—may be well illustrated by taking the case of light. Illuminating power is dependent, first on the extent of the light-giving source. If we double a gas flame we get double the amount of light. Two candles together will give twice the light of one of them at the same distance, and will cast twice as strong a shadow. But again we can see to read as clearly with one candle as with two, if the single flame be brought nearer than the double flame. And one candle flame a yard off will give us the light not of two, but of *four* similar flames two yards away; so that a decrease of distance more than compensates for a decrease of flame at the same rate. The reason of this is manifest from the following illustration :—

What proves the Geometrical Law of areas, is supposed to prove that the ray leaves its straight line and that Light spreads.—"A board a foot square, represented by A B, placed at any distance from a candle at C, will just shadow a board, E D, of two feet square placed at double the distance, and one of three feet square, L K, placed at triple the distance. But E D will have *four* times as much surface as A B, because the former is both twice as long and twice as broad as the latter; and the board L K of three feet square, will in like manner have nine times as much surface as A B. Thus the light that A B would catch will be SPREAD OVER four times as much space at E D, nine times at L K ; and CONSEQUENTLY, it is only one-fourth as strong at double the distance, one-ninth at triple the distance, one-sixteenth at four times, and so on.

" So that if we had a bell ringing at C (*instead of a light there*) the amount of sound that would be caught by an ear-trumpet with an opening of, say, one square foot, placed at A B, will have spread out at double the distance (E D) over four times the space ; and the ear-trumpet there would catch only one-fourth of the sound it caught at B, and therefore the sound would be only one-fourth as strong.

" In more technical language the law is expressed, ' The intensity varies inversely as the square of the distance ; that is to say, the intensity of light, sound, gravitation, &c., increases or decreases at the square of the rate that the distance decreases or increases ' (p. 8, 9).

" Sound, like gravitation, *light*, heat, or any other uniformly *spreading* influence, follows the law of the intensity, &c." (p. 324).

Illustration by Boxes and Tapers instead of Concentric Spheres.—" Light, like any other emanation from a central point, in *spreading* through wider space, becomes thinner or less intense in proportion as *it spreads*. Thus, if a taper be placed in the centre of a cubical box, every side of which is a foot square, the light falling on the sides of the box will have a certain intensity there ; if the taper be then placed in a similar box with sides of two feet square, there will be only the same quantity of light, but it will be *spread over* four times as much surface (for a square having two feet in the edge is made up of four squares of one foot), and will therefore, *on any part* of that surface, be only one-fouth part as strong or intense as in the first box; and so for any other size of box or space, the intensity will diminish as the square of the distance increases.

Law of the Squares supposed independent of Enlarged Space. —"Hence if the earth were at twice its present distance from the sun, *i.e.*, beyond the planet Mars, it would receive only one-fourth of the light and heat which it now receives, just as a man placed four yards from a fire receives only one-fourth of the heat which falls on a man at two yards. At three times its present distance from the sun, the earth would receive only one-ninth of the light now received, and conversely, if it were only one-third of the present distance, the heat and light would be increased ninefold " (p. 561–2).

In Dr. Arnott's own edition of 1864, I find the following passages :—

"Now, *light*, heat, attraction, sound, and indeed every influence *spreading*, uniformly *from a central point*, is found to decrease in the proportions here illustrated, viz., as the surface of squares,WHICH SHADOW ONE ANOTHER, increases. The technical expression is, 'the intensity is inversely as the square of the distance,'—the distances being estimated from the centre of attraction or radiation " (p. 13).

" It is to be remarked that no substances are absolutely opaque, and *none perfectly transparent.* . . . The purest water and air arrest a part of the light which enters them. A depth of 20 feet of pure water intercepts nearly half of the light which enters it; and of the sun's light when passing through a great extent of atmosphere, as shortly after sunrise or before sunset, *a considerable portion* is absorbed and lost " (p. 507).

" It has already been explained that light, like every other influence radiating from a centre, becomes *rapidly* weaker as the distance from the centre increases,—being, for instance, only one-fourth part as intense at double distance,—*while* it is *still farther* weakened by *the obstacle of any transparent medium* through which it passes " (p. 562).

No. 16.

Translated from Ganot's "Natural Philosophy."

" All bodies which transmit Light extinguish or absorb a portion of the Light. The most transparent, such as air, water, glass, gradually extinguish the Light which penetrates them; and if their thickness (*i.e.*, width or length) be considerable, they may weaken it so much that no impression is produced on the eye."

" The atmosphere has everywhere the same composition, but not the same density, owing to the variations in pressure and temperature to which it is subject in various places."

" Light emanates from luminous bodies, in ALL directions; for we see them equally in ALL positions in which we are placed around them. . . . This emanation of Light in all directions about a luminous body is called Radiation. A luminous ray, or ray of Light, is the line in which Light is propagated. A luminous pencil, or pencil of Light, is a collection of rays from the same source. It is said to be parallel when it is composed of parallel rays;—divergent, when the rays *separate from each other;* and convergent, when they tend towards the same point."

" The intensity of a source of Light is measured by the quantity of Light which it sends on a given surface. From the property which rays have of diverging, this quantity of Light, this intensity, decreases rapidly as the illuminated body is removed from the luminous body. It may be shown by geometrical considerations that the intensity of Light is (thus) inversely as the square of the distance."

" It is important to observe that it is *in consequence of the divergence* of luminous rays that Light decreases as the distance increases. This decrease does not obtain in the case of parallel rays; their lustre would be the same AT ALL DISTANCES, were it not for the absorption which takes place in even the most diaphanous media."

No. 17.

Faraday, in his admirable little work on the " Various Forces of Nature," says :—

"Now I want you clearly to understand what this law is. They say (and they are right) that two bodies attract each other *inversely as the square of the distance*,—a sad jumble of words until you understand them; but I think we shall soon comprehend what this law is, and what is the meaning of the 'inverse square of the distance.' I have here a lamp *A* shining most intensely upon this disc *B C D*, and this light acts as a sun by which I can get a shadow from this little screen *B F* (merely a square piece of card) which, as you know, when I place it close to the large screen, just shadows as much of it as is exactly equal to its own size. But now let me take this card E, which is equal to the other one in size, and place it midway between the lamp and the screen. Now look at the size of the shadow B D;—it is four times the original size. Here then comes the 'inverse square of the distance.' This distance A E is *one*, and that distance A B, is *two;* but that size E being *one*, this size B D of shadow is *four* instead of two, which is the *square* of the distance; and if I put the screen at one-third of the distance from the lamp, the shadow on the large screen would be nine times the size. Again, if I hold this screen *here* at B F, a certain amount of light falls on it; and if I hold it nearer the lamp at *E*, *more* light shines upon it. And you see at once how much,— exactly the quantity which I have shut off from the part of this screen, B D, now in shadow; moreover you see that, if I put a single screen here, at G by the side of the shadow, it can only receive *one-fourth* of the proportion of light which is obstructed. That then is what is meant by the *inverse* of the square of the distance. This screen is the brightest because it is the nearest; and there is the whole secret of this curious expression, 'inversely as the square of the distance.' Now if you cannot perfectly recollect this when you go home, get a candle, &c., &c." (pp. 46-8).

No. 18.

Helmholtz, the enlightened and distinguished Professor of Natural Philosophy at the University of Berlin, writes as follows, most nearly approaching, in candour and clearness, the celebrated French writer whose labours he adverts to :—

" The common theory supposes that there is no loss of Light at all in the cosmical space, but that all the light of the sun which reaches the distance of one million of miles reaches also the distance of two millions, and spreads out in the latter case over an area 4 times as great as in the first case. If there should be any absorption of Light,— any loss of its quantity in the cosmical space,—then, of course, the diminution of its intensity with increasing distance would be far greater. I don't see any cause why we should suppose that any absorption takes place (there). For terrestrial distances the law of the Inverse Squares has been proved experimentally at first by Bouguer" (March, 1876).

No. 19.

Schiaparelli, the distinguished Italian astronomer, Director of the Royal Observatory at the Brera in Milan, makes the following observation, in a long letter on the Inverse Squares, and on the confusion made between that theory and the diminution in a Medium :—

" As far as I am concerned, I shall be but too happy if my remarks, already carried to too great a length, have the

effect of leading you to adduce further considerations calculated to throw some light on this difficult subject" (Sep-
tember, 1874).   The original words are :—

Queste ed altre ragioni forse anche più cogenti saranno
probabilmente quelle, che hanno indotti i fisici illustri da
V. S. citati a pronunziarsi in modo dubitativo circa il
significato che Ella attribuisce alla legge dell' inversa dei
quadrati.   Per conto mio io sarò contento, se questa già
troppo lunga diceria Le darà occasione di produrre sul
medesimo argomento nuove ragioni a sua difesa, che
valgano ad illustrare la scabrosa materia.

No. 20.

Dr. Thomas Young in his " Lectures on Natural Philosophy," has the following passages :—

" We cannot exhibit a single ray of light except as the
confine between light and *darkness*, or as the lateral limit of
a pencil of light.

" Perhaps no medium is strictly speaking absolutely
transparent; for, even in the air, a *considerable* portion of
light is intercepted.   It has been estimated that of the horizontal sunbeams passing through about 200 miles of air, *one
two-thousandth part* only reaches us, and that no *sensible*
light can penetrate more than seven hundred feet deep into
the sea; a length of seven feet of water having been
found to intercept *one-half of the light* which enters it."
Sec. 35.)

For this purpose (to compare the intensity of the light
afforded by any two luminous objects) it is necessary to
assume as a principle that the *same* quantity, diverging in
all directions from a luminous body, *remains undiminished at
all distances* from the centre of divergence.   Thus we *must*
suppose that the quantity of light falling on every body is
the same as would have fallen on the place occupied by its

shadow ; and if there were any doubt of the truth of the *supposition*, it might be confirmed by some simple experiments. *It follows* that since the shadow of a square inch of any surface, occupies, at twice the distance of the surface from the luminous point, the space of four square inches, *this intensity of the light diminishes as the square of the distances increases* " (Sec. 36).

"It appears that luminous bodies in general emit light equally in every direction, *not from each point* of any of their surfaces, *as some have supposed*, but from the *whole* surface *taken together*, so that the surface when viewed obliquely, appears neither more nor less bright than when viewed directly " (Sec. 37).

" The law of gravitation, which indicates the ratio of its increase with the diminution of the distance, is principally deduced from astronomical observations and computations ; it is the simplest that can be conceived for any influence, that either spreads from a centre or converges towards a centre ; for it supposes the force acting on the same substance, to be always proportional to the *angular* space that it occupies " (Sec. 49).

END OF APPENDIX.

INDEX.

INDEX.

END OF INDEX.

McCorquodale & Co., Printers, "The Armoury," Southwark.

WILLIAMS AND NORGATE'S

LIST OF

French, German, Italian, Latin, and Greek,

AND OTHER

SCHOOL BOOKS AND MAPS.

French.

FOR PUBLIC SCHOOLS WHERE LATIN IS TAUGHT.

Eugène (G.) The Student's Comparative Grammar of the French Language, with an Historical Sketch of the Formation of French. For the use of Public Schools. With Exercises. 2nd Improved Edition. Square crown 8vo, cloth. 5*s.*

Or Grammar, 3*s.* ; Exercises, 2*s.* 6*d.*

"The appearance of a Grammar like this is in itself a sign that great advance is being made in the teaching of modern as well as of ancient languages. . . . . The rules and observations are all scientifically classified and explained. . . . . Mr. Eugène's book is one that we can strongly recommend for use in the higher forms of large schools."—*Educational Times.*

"In itself this is in many ways the most satisfactory Grammar for beginners that we have as yet seen. . . . . The book is likely to be useful to all who wish either to learn or to teach the French language."—*Athenæum.*

Eugène's French Method. Elementary French Lessons. Easy Rules and Exercises preparatory to the "Student's Comparative French Grammar." By the same Author. 2nd Edition. Crown 8vo, cloth. 1*s.* 6*d.*

"Certainly deserves to rank among the best of our Elementary French Exercise-books."—*Educational Times.*

"To those who begin to study French, I may recommend, as the best book of the kind with which I am acquainted, '*Eugène's Elementary Lessons in French.*' It is only after having fully mastered this small manual and Exercise-book that they ought to begin the more systematic study of French."—*Dr. Breymann, Lecturer of the French Language and Literature, Owen's College, Manchester (Preface to Philological French Grammar).*

Eugène's Comparative French-English Studies, Grammatical and Idiomatic. Being a Second, entirely re-written, Edition of the "French Exercises for Middle and Upper Forms." Cloth. 2*s.* 6*d.*

Attwell (H.) Twenty Supplementary French Lessons, with Etymological Vocabularies. Chiefly for the use of Schools
where Latin is taught. Crown 8vo, cloth. 2*s.*

Krueger (H.) Short French Grammar. 4th Edition. 180 pp.
12mo, cloth. 2*s.*

Eugène (G.) French Irregular Verbs scientifically classified with
constant Reference to Latin. Reprinted from his Gram-
mar. 8vo, sewed. 6*d.*

Ahn's French Familiar Dialogues, and French-English Vocabulary for English Schools. 12mo, cloth. 2*s.*

Brasseur (Prof. Isid.) Grammar of the French Language, comprehending New and complete Rules on the Genders of
French Nouns. 20th Edition. 12mo, cloth. 3*s.* 6*d.*

—— Key to the French Grammar. 12mo, cloth. 3*s.*

—— Selection from Chesterfield's and Cowper's Letters, with
Notes for translating. 5th Edition. 12mo, cloth. 3*s.*

—— Key. Partie Française du Choix des Lettres. 12mo,
cloth. 3*s.* 6*d.*

—— Manuel des Ecoliers. A French Reading Book, preceded by Rules on French Pronunciation. 6th Edition.
12mo. 2*s.* 6*d.*

—— Premières Lectures. An easy French Reading Book
for Children and Beginners. 18mo, cloth. 1*s.* 6*d.*

Roche (A.) Nouvelle Grammaire Française. Nouvelle Edition.
12mo, boards. 1*s.*

Williams (T. S.) and J. Lafont. French and Commercial Correspondence. A Collection of Modern Mercantile Letters
in French and English, with their translation on opposite
pages. 2nd Edition. 12mo, cloth. 4*s.* 6*d.*

For a German Version of the same Letters, vide p. 4.

Fleury's Histoire de France, racontée à la Jeunesse, edited for
the use of English Pupils, with Grammatical Notes, by
Auguste Beljame, Bachelier-ès-lettres de l'Université de
Paris. 2nd Edition. 12mo, cloth boards. 3*s.* 6*d.*

Mandrou (A.) Album Poétique de la Jeunesse. A Collection of
French Poetry, selected expressly for English Schools by
A. Mandrou, M.A. de l'Académie de Paris, Professor of
French in the Clergy Orphan School, St. Peter's Colle-
giate School, the Crystal Palace, &c. 12mo, cloth. 3*s.* 6*d.*

German.

Weisse's Complete Practical Grammar of the German Language, with Exercises on Conversations, Letters, Poems and Treatises, &c. 3rd Edition, very much improved. 12mo, cloth.   *6s.*

——— **New Conversational Exercises in German Composition,** with complete Rules and Directions, with full References to his German Grammar. 2nd Edition. 12mo, cloth.   *3s. 6d.*

Schlutter's German Class Book. A Course of Instruction based on Becker's System, and so arranged as to exhibit the Self-development of the Language, and its Affinities with the English. By Fr. Schlutter, Royal Military Academy, Woolwich. 3rd Edition. 12mo, cloth.   *5s.*

Möller (A.) A German Reading Book. A Companion to Schlutter's German Class Book. With a complete Vocabulary. 150 pp. 12mo, cloth.   *2s.*

Wittich's German Grammar. 7th Edition. 12mo, cloth. *6s. 6d.*

——— **German for Beginners.** New Edition. 12mo, cloth. *5s.*

——— **Key to ditto.** 12mo, cloth.   *7s.*

——— **German Tales for Beginners,** arranged in Progressive Order. 20th Edition. Crown 8vo, cloth.   *6s.*

Ravensberg (A. v.) Practical Grammar of the German Language. Conversational Exercises, Dialogues and Idiomatic Expressions. 2 vols. in 1. 12mo, cloth.   *5s.*

——— **Key to the Exercises.** Cloth.   *2s.*

——— **Rose's English into German.** A Selection of Anecdotes, Stories, Portions of Comedies, &c., with copious Notes for Translation into English. By A. v. Ravensberg. 2nd Edition. 2 Parts in 1. Cloth.   *4s. 6d.*

——— **Key to Rose's English into German.** Cloth.   *5s.*

——— **German Reader,** Prose and Poetry, with copious Notes for Beginners. 2nd Edition. Crown 8vo, cloth.   *3s.*

——— **Student's First Year's German Companion.** A concise Conversational Method for Beginners. 12mo, cloth.   *2s. 6d.*

Sonnenschein and Stallybrass. German for the English. Part I. First Reading Book. Easy Poems with interlinear Translations, and illustrated by Notes and Tables, chiefly Etymological. 4th Edition. 12mo, cloth.   *4s. 6d.*

Ahn's German Method by Rose. A New Edition of the genuine Book, with a Supplement consisting of Models of Conjugations, a Table of all Regular Dissonant and Irregular Verbs, Rules on the Prepositions, &c. &c. By A. V. Rose. 2 Courses in 1 vol. Cloth.   3*s.* 6*d.*

—————— **German Method by Rose, &c.** First Course. Cloth. 2*s.*

Apel's Short and Practical German Grammar for Beginners, with copious Examples and Exercises. 2nd Edition. 12mo, cloth.   2*s.* 6*d.*

[Black's] Thieme's Complete Grammatical German Dictionary, in which are introduced the Genitives and Plurals and other . Irregularities of Substantives, the Comparative Degrees of Adjectives, and the Irregularities of Verbs. Square 8vo, strongly bound.   6*s.*

Koehler (F.) German-English and English-German Dictionary. 2 vols. 1120 pp., treble columns, royal 8vo, in one vol., half-bound.   9*s.*

Williams (T. S.) Modern German and English Conversations and Elementary Phrases, the German revised and corrected by A. Kokemueller. 21st enlarged and improved Edition. 12mo, cloth.   3*s.* 6*d.*

—————— **and O. Cruse. German and English Commercial Correspondence.** A Collection of Modern Mercantile Letters in German and English, with their Translation on opposite pages. 12mo, cloth.   4*s.* 6*d.*

For a French Version of the same Letters, vide p. 2.

Apel (M.) German Prose Stories for Beginners (including Lessing's Prose Fables), with an interlinear Translation in the natural order of Construction. 12mo, cloth. 2*s.* 6*d.*

—————— **German Poetry.** A Collection of German Poetry for the use of Schools and Families, containing nearly 300 Pieces selected from the Works of 70 different Authors. Crown 8vo, cloth.   5*s.*

—————— **German Prose.** A Collection of the best Specimens of German Prose, chiefly from Modern Authors. A Handbook for Schools and Families. 500 pp. Crown 8vo, cloth.   3*s.*

Andersen (H. C.) Bilderbuch Ohne Bilder. The German Text, with Explanatory Notes, &c., and a complete Vocabulary, for the use of Schools, by Alphons Beck. 12mo, cloth limp.   2*s.*

Chamisso's Peter Schlemihl. The German Text, with copious Explanatory Notes and a Vocabulary, by M. Förster. Crown 8vo, cloth. 2*s.*

Goethe's Hermann und Dorothea. With Grammatical and Explanatory Notes and a complete Vocabulary, by M. Förster. 12mo, cloth. 2*s.* 6*d.*

———— Hermann und Dorothea. With Grammatical Notes by A. von Ravensberg. Crown 8vo, cloth. 2*s.* 6*d.*

———— Hermann und Dorothea. The German Text, with corresponding English Hexameters on opposite pages. By F. B. Watkins, M.A., Professor of Greek and Latin, Queen's College, Liverpool. Crown 8vo, cloth. 3*s.*

———— Egmont. The German Text, with Explanatory Notes and a complete Vocabulary, by H. Apel. 12mo, cloth. 2*s.* 6*d.*

———— Faust. With copious Notes by Falk Lebahn. 8vo, cloth. 10*s.* 6*d.*

Goldschmidt (H. E.) German Poetry. A Selection of the best Modern Poems, with the best English Translations on opposite pages. Crown 8vo, cloth. 5*s.*

Hauff's Mærchen. A Selection from Hauff's Fairy Tales. The German Text, with a Vocabulary in foot-notes. By A. Hoare, B.A. Crown 8vo, cloth. 3*s.* 6*d.*

Nieritz. Die Waise, a German Tale, with numerous Explanatory Notes for Beginners, and a complete Vocabulary, by E. C. Otte. 12mo, cloth. 2*s.* 6*d.*

Carové (J. W.) Mæhrchen ohne Ende (The Story without an End). 12mo, cloth. 2*s.*

Lessing's Minna von Barnhelm, the German Text, with Explanatory Notes for translating into English, and a complete Vocabulary, by J. A. F. Schmidt. 12mo, cloth. 2*s.* 6*d.*

Schiller's Song of the Bell, German Text, with English Poetical Translation on the opposite pages, by J. Hermann Merivale, Esq. 12mo, cloth. 1*s.*

Fouque's Undine, Sintram, Aslauga's Ritter, die beiden Hauptleute. 4 vols. in 1. 8vo, cloth. 7*s.* 6*d.*

Undine. 1*s.* 6*d.*; cloth, 2*s.*   Aslauga. 1*s.* 6*d.*; cloth, 2*s.*
Sintram. 2*s.* 6*d.*; cloth, 3*s.*   Hauptleute. 1*s.* 6*d.*; cloth, 2*s.*

Latin and Greek.

Jessopp (Rev. Dr.) Manual of Greek Accidence. New Edition.
Crown 8vo.   3s. 6d.

Bryce (Rev. Dr.) The Laws of Greek Accentuation Simplified.
3rd Edition, with the most essential Rules of Quantity.
12mo, sewed.   6d.

Euripides' Medea. The Greek Text, with Introduction and
Explanatory Notes for Schools, by J. H. Hogan. 8vo,
cloth.   3s. 6d.

————— Ion. Greek Text, with Notes for Beginners, Introduction and Questions for Examination, by the Rev. Charles
Badham, D.D. 2nd Edition. 8vo.   3s. 6d.

Æschylus. Agamemnon. Revised Greek Text, with literal
line-for-line Translation on opposite pages, by John F.
Davies, B.A. 8vo, cloth.   3s.

Platonis Philebus. With Introduction and Notes by Dr. C.
Badham. 2nd Edition, considerably augmented. 8vo,
cloth.   4s.

————— Euthydemus et Laches. With Critical Notes and an
Epistola critica to the Senate of the Leyden University,
by the Rev. Ch. Badham, D.D. 8vo, cloth.   4s.

————— Symposium, and Letter to the Master of Trinity, "De
Platonis Legibus,"—Platonis Convivium, cum Epistola
ad Thompsonum edidit Carolus Badham. 8vo, cloth. 4s.

Sophocles. Electra. The Greek Text critically revised, with
the aid of MSS. newly collated and explained. By Rev.
H. F. M. Blaydes, M.A., formerly Student of Christ
Church, Oxford. 8vo, cloth.   6s.

————— Philoctetes. Edited by the same. 8vo, cloth.   6s.

————— Trachiniæ. Edited by the same. 8vo, cloth.   6s.

————— Ajax. Edited by the same. 8vo, cloth.   6s.

Kiepert's New Atlas Antiquus. Maps of the Ancient World,
for Schools and Colleges. 6th Edition. With a complete Geographical Index. Folio, boards.   7s. 6d.

————— The same, without the Index.   6s. 6d.

Italian.

Volpe (Cav. G.) Eton Italian Grammar, for the use of Eton College. Including Exercises and Examples. New Edition. Crown 8vo, cloth. 4*s.* 6*d.*
———— Key to the Exercises. 1*s.*
Rossetti. Exercises for securing Idiomatic Italian by means of Literal Translations from the English, by Maria F. Rossetti. 12mo, cloth. 3*s.* 6*d.*
———— Aneddoti Italiani. One Hundred Italian Anecdotes, selected from " Il Compagno del Passeggio." Being also a Key to Rossetti's Exercises. 12mo, cloth. 2*s.* 6*d.*

Danish—Dutch.

Bojesen (Mad. Marie) The Danish Speaker. Pronunciation of the Danish Language, Vocabulary, Dialogues and Idioms for the use of Students and Travellers in Denmark and Norway. 12mo, cloth. 4*s.*
Rask (E.) Danish Grammar for Englishmen. With Extracts in Prose and Verse. 2nd Edition. Edited by Repp. 8vo. 5*s.*
Ferrall, Repp, and Rosing. Danish-English and English-Danish Dictionary. New Edition. 2 Parts in 1. Square 8vo. 14*s.*
Williams and Ludolph. Dutch and English Dialogues, and Elementary Phrases. 12mo. 2*s.* 6*d.*

Wall Maps.

Sydow's Wall Maps of Physical Geography for School-rooms, representing the purely physical proportions of the Globe, drawn on a very large scale. An English Edition, the Originals with English Names and Explanations. Mounted on canvas, with rollers :

1. The World. 12 Sheets. Mounted. 10*s.*
2. Europe. 9 Sheets. Mounted. 10*s.*
3. Asia. 9 Sheets. Mounted. 10*s.*
4. Africa. 6 Sheets. 10*s.*
5. America (North and South). 2 Maps, 10 Sheets. 10*s.*
6. Australia and Australasia. 6 Sheets. Mounted. 10*s.*
———— Handbook to the Series of Large Physical Maps for School Instruction, edited by J. Tilleard. 8vo. 1*s.*

Miscellaneous.

De Rheims (H.). Practical Lines in Geometrical Drawing, containing the Use of Mathematical Instruments and the Construction of Scales, the Elements of Practical and Descriptive Geometry, Orthographic and Horizontal Projections, Isometrical Drawing and Perspective. Illustrated with 300 Diagrams, and giving (by analogy) the solution of every Question proposed at the Competitive Examinations for the Army. 8vo, cloth.　　9*s.*

Fuerst's Hebrew Lexicon, by Davidson. A Hebrew and Chaldee Lexicon to the Old Testament, by Dr. Julius Fuerst. 4th Edition, improved and enlarged, containing a Grammatical and Analytical Appendix. Translated by Rev. Dr. Samuel Davidson. 1600 pp., royal 8vo, cloth. 21*s.*

Hebrew Texts. Large type. 16mo, cloth.　　each 1*s.*
　The Book of Genesis.　　1*s.*
　The Psalms.　　1*s.*
　The Book of Job.　　1*s.*
　The Prophet Isaiah.　　1*s.*

Attwell (Prof. H.) Table of Aryan (Indo-European) Languages, showing their Classification and Affinities, with copious Notes; to which is added, Grimm's Law of the Interchange of Mute Consonants, with numerous Illustrations. A Wall Map for the use of Colleges and Lecture-rooms. 2nd Edition. Mounted with rollers.　　10*s.*

—— The same Table, in 4to, with numerous Additions. Boards.　　7*s.* 6*d.*

Williams and Simmonds. English Commercial Correspondence. A Collection of Modern Mercantile Letters. By T. S. Williams and P. L. Simmonds, Author of "A Dictionary of Trade Products," Editor of "The Technologist." 12mo, cloth.　　4*s.*

Bayldon. Icelandic Grammar. An Elementary Grammar of the Old Norse or Icelandic Language. By Rev. George Bayldon. 8vo, cloth.　　7*s.* 6*d.*

Small's Handbook of Sanskrit Literature. Compiled from the best Authorities by the Rev. G. Small, formerly Missionary at Calcutta and Benares. Intended specially for the use of Candidates for the Indian Civil Service. Crown 8vo, cloth.　　6*s.*